普通高等教育"十三五"规划教材（软件工程专业）

Java 程序设计实训教程

主　编　宁淑荣　杨国兴

副主编　金忠伟　李田英

中国水利水电出版社
www.waterpub.com.cn

·北京·

内 容 提 要

本书精心设计了4个实训：扫雷游戏、网络五子棋、棋谱的保存与回放、学生成绩管理系统，介绍 Java 在应用软件开发中用到的主要技术，并体现面向对象的设计思想。对于 Java 中的输入输出、数据库、异常处理、网络编程、界面编程等都有比较深入的训练。

本书可作为计算机类专业 Java 实训、Java 课程设计等课程的教材，也可作为学生毕业设计以及 Java 程序设计爱好者的参考书。

本书配有电子教案和源代码，读者可以到中国水利水电出版社网站和万水书苑上免费下载，网址为 http://www.waterpub.com.cn/softdown 和 http://www.wsbookshow.com。

图书在版编目（CIP）数据

Java程序设计实训教程 / 宁淑荣，杨国兴主编. -- 北京：中国水利水电出版社，2018.1
普通高等教育"十三五"规划教材. 软件工程专业
ISBN 978-7-5170-6121-2

Ⅰ. ①J… Ⅱ. ①宁… ②杨… Ⅲ. ①JAVA语言－程序设计－高等学校－教材 Ⅳ. ①TP312

中国版本图书馆CIP数据核字(2017)第302641号

策划编辑：周益丹　　责任编辑：封 裕　　封面设计：李 佳

书　名	普通高等教育"十三五"规划教材（软件工程专业） Java 程序设计实训教程 Java CHENGXU SHEJI SHIXUN JIAOCHENG
作　者	主　编　宁淑荣　杨国兴 副主编　金忠伟　李田英
出版发行	中国水利水电出版社 （北京市海淀区玉渊潭南路1号D座　100038） 网址：www.waterpub.com.cn E-mail：mchannel@263.net（万水） 　　　　sales@waterpub.com.cn 电话：（010）68367658（营销中心）、82562819（万水）
经　售	全国各地新华书店和相关出版物销售网点
排　版	北京万水电子信息有限公司
印　刷	三河市铭浩彩色印装有限公司
规　格	184mm×260mm　16开本　11.25印张　271千字
版　次	2018年1月第1版　2018年1月第1次印刷
印　数	0001—3000册
定　价	26.00元

凡购买我社图书，如有缺页、倒页、脱页的，本社营销中心负责调换
版权所有·侵权必究

前　　言

　　Java 是目前使用最广泛的语言之一。对于软件开发人员来说，掌握 Java 语言基础并拥有使用 Java 进行软件开发的能力是非常重要的，因此大多数与计算机相关的专业都开设了 Java 程序设计课程。

　　Java 程序设计是一门实践性很强的课程（任何一种计算机语言课程皆是），仅仅掌握 Java 的基本语法知识，与能利用 Java 进行软件开发还有很大的差距。掌握 Java 基本知识后，应该通过大量的编程实践来逐步提高利用 Java 进行软件开发的能力。本书精心设计了 4 个实训，详细介绍具体的开发过程，读者可以跟随书中介绍的步骤轻松完成实训程序的开发。

　　本书由 4 个实训组成：扫雷游戏、网络五子棋、棋谱的保存与回放、学生成绩管理系统，涉及的主要知识有异常处理、输入输出流、数据库编程、多线程和网络编程等。

　　本书中的所有程序都由作者亲自编写，并在 JDK1.6 环境下调试通过，实例中用到的数据库是 MySQL 数据库。

　　为了方便教师教学和学生学习，本书提供用 PowerPoint 制作的电子教案，教师可根据具体情况进行必要的修改后使用。

　　本书由宁淑荣、杨国兴任主编，金忠伟、李田英任副主编，其中实训 1 至实训 3 由宁淑荣编写，实训 4 由杨国兴、金忠伟和李田英共同编写。

　　本书的编写和出版得到了"北京联合大学规划教材建设项目"资助，特此感谢。

　　由于作者水平有限，书中不妥之处在所难免，恳请专家与读者批评指正。

编　者
2017 年 10 月

目　　录

前言

实训1　扫雷游戏 1
1.1　系统设计 1
1.1.1　需求分析 1
1.1.2　类的设计 3
1.2　创建主窗口 4
1.2.1　创建项目 5
1.2.2　MineFrame 类 5
1.2.3　主程序类 Saolei 7
1.3　MinePanel 类和 Block 类 8
1.3.1　几个辅助类 8
1.3.2　Block 类 9
1.3.3　MinePanel 类 12
1.3.4　将雷区加入到 MineFrame 中 14
1.4　实现扫雷功能 16
1.4.1　翻开小方块 16
1.4.2　处理输赢以及搜索方法 20
1.4.3　加快扫雷进程 22
1.4.4　重新开始游戏 23
1.5　选择游戏难度级别 23
1.5.1　在 MineFrame 类中添加 grade 属性 23
1.5.2　自定义难度对话框 24
1.5.3　完善菜单监听器类 26
1.6　实现计时功能 27
1.6.1　UpdateTimeTask 类 27
1.6.2　启动计时与终止计时 27
1.7　扫雷排行榜 28
1.7.1　Record 类 28
1.7.2　RecordDao 类 30
1.7.3　用于输入游戏者名字的对话框类 31
1.7.4　显示排行榜的对话框类 32
1.7.5　实现排行榜功能 34
1.8　附加功能 36
1.8.1　添加 sound()方法 36
1.8.2　准备音频文件 37
1.8.3　播放音频文件 37
1.9　作业 37

实训2　网络五子棋 38
2.1　单机版五子棋游戏 39
2.1.1　五子棋游戏窗口制作 40
2.1.2　创建棋盘类 41
2.1.3　创建棋子类 42
2.1.4　实现单击鼠标下棋 44
2.1.5　判断赢棋 46
2.1.6　实现工具栏上的功能 49
2.1.7　改变鼠标的形状 50
2.2　服务器端界面制作 51
2.3　创建客户端界面 52
2.3.1　创建主窗口和棋盘 52
2.3.2　创建客户端界面右侧的三个类 53
2.3.3　创建客户端界面下方的控制面板类 56
2.4　实现"连接主机"按钮的功能 56
2.4.1　连接服务器获取用户名 57
2.4.2　将已经连接的客户端添加到用户列表中 60
2.5　实现"加入游戏"按钮的功能 64
2.5.1　客户端申请加入后对方选择同意或拒绝 65
2.5.2　完成猜棋并准备好下棋 70
2.6　实现下棋功能 73
2.6.1　客户端向服务器发送下棋消息 74
2.6.2　服务器接收消息并处理 75
2.6.3　客户端接收消息并处理 76
2.7　实现"放弃游戏"按钮的功能 78
2.7.1　Command 类添加常量 78

2.7.2　添加"放弃游戏"按钮的响应代码…78
　　2.7.3　在 Communication 类中添加
　　　　　giveup()方法………………………78
　　2.7.4　服务器接收 giveup 命令并处理…79
2.8　加入计时功能……………………………79
　　2.8.1　设计计时线程类……………………79
　　2.8.2　猜先后启动倒计时线程……………80
2.9　完善"关闭程序"按钮的功能…………81
　　2.9.1　在 Command 类中添加命令………81
　　2.9.2　客户端向服务器发送命令…………81
　　2.9.3　服务器处理 quit 命令………………82
　　2.9.4　客户端处理 delete 命令……………82
2.10　作业……………………………………82

实训 3　棋谱的保存与回放………………84
3.1　创建数据库………………………………84
　　3.1.1　数据库设计…………………………84
　　3.1.2　数据库创建…………………………85
3.2　用户管理…………………………………87
　　3.2.1　数据库连接类………………………87
　　3.2.2　用户管理……………………………88
3.3　用户注册和登录…………………………92
　　3.3.1　准备工作……………………………93
　　3.3.2　用户登录……………………………94
　　3.3.3　用户注册……………………………98
3.4　记录棋局和棋谱…………………………101
　　3.4.1　记录棋局……………………………102
　　3.4.2　记录棋谱……………………………108
3.5　查询棋局和棋谱欣赏……………………111
　　3.5.1　查询棋局……………………………111

　　3.5.2　棋谱欣赏……………………………115
3.6　作业………………………………………119

实训 4　学生成绩管理系统………………120
4.1　系统设计…………………………………120
　　4.1.1　需求分析……………………………120
　　4.1.2　数据库设计…………………………124
　　4.1.3　类的设计……………………………124
4.2　工具类……………………………………125
　　4.2.1　DBConnection 类……………………125
　　4.2.2　CreateDatabase 类…………………127
4.3　实体类……………………………………129
　　4.3.1　班级实体类 ClassEntity……………129
　　4.3.2　学生实体类 Student…………………129
　　4.3.3　课程实体类 Course…………………131
　　4.3.4　成绩实体类 Score……………………132
4.4　数据访问类………………………………133
　　4.4.1　ClassDao 类…………………………133
　　4.4.2　StudentDao 类………………………136
　　4.4.3　CourseDao 类…………………………139
　　4.4.4　ScoreDao 类…………………………141
4.5　主窗口类…………………………………144
4.6　班级管理…………………………………146
4.7　学生管理…………………………………152
4.8　成绩管理…………………………………159
　　4.8.1　准备工作……………………………159
　　4.8.2　成绩录入与修改……………………162
　　4.8.3　成绩查询……………………………167
4.9　作业………………………………………171

参考文献……………………………………172

实训 1　扫雷游戏

使用 Java 语言编写一个类似 Windows 系统中的扫雷游戏，游戏界面由若干行、若干列的小方块组成，在这些小方块中随机布置着若干个雷，游戏的任务就是将其中的雷找到，将不是雷的小方块翻开，若将所有不是雷的小方块成功翻开则游戏成功，若翻开是雷的小方块则游戏失败。运行界面如图 1.1 所示。

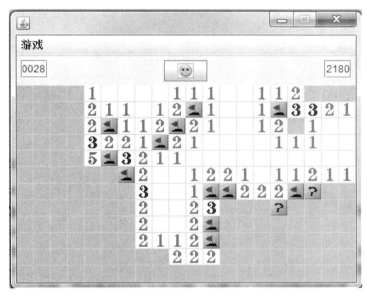

图 1.1　扫雷游戏界面

游戏窗口包含菜单"游戏"，可通过菜单选择游戏的难度级别；菜单的下一行包括三部分，单击中间的按钮可以重新开始游戏，左侧的文本框显示剩余的雷数（没有标记出来的雷数）；右侧的文本框显示游戏已经使用的时间（秒）。

本实训首先对扫雷游戏进行分析设计，然后逐步实现扫雷游戏的所有功能。

1.1　系统设计

1.1.1　需求分析

1. 随机布雷

游戏开始前，在雷区随机布雷，也就是将若干个雷随机分布在雷区的小方块下面。

2. 实现扫雷功能

在小方块上单击鼠标左键，如果该小方块下面是雷，则游戏失败，并将所有雷显示出来；如果该小方块下面不是雷，则将其翻开，翻开后显示的数字是其周围的雷数，每个小方块周围有 8 个相邻的小方块，因此显示的数字应在 0~8 之间，如果该小方块周围的雷数是 0，则自动将其周围的小方块也翻开。

3. 标记小方块

如果游戏者能够确定某个小方块是雷，可以单击鼠标右键，将其标记为雷（标记为小红旗），在标记为小红旗的小方块上再次单击鼠标右键可将其标记为问号，表示怀疑该小方块是雷，在标记为问号的小方块上再次单击鼠标右键则恢复为原始状态。

4. 选择游戏的难度

扫雷程序分为三个级别：初级、中级和高级，级别越高雷区越大而且雷越多。通过菜单可以选择游戏的级别，如图 1.2 所示；还可以通过"自定义"菜单项自由设置雷区的行数、列数和雷数。

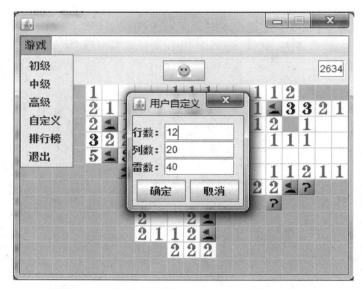

图 1.2　扫雷"游戏"菜单和"用户自定义"对话框

5. 显示剩余雷数和已经使用的时间

在窗口上方两侧的两个文本框中分别显示未标记的雷数和游戏已经使用的时间。

6. 最好成绩纪录

将扫雷成功的最少时间（每个级别分别记录）记录在文件中，每当扫雷成功后，将游戏用时与原来的纪录比较，如果当前使用的时间较少，则保存新的时间作为纪录。在保存成绩之前，要通过对话框输入游戏者的名字，如图 1.3 所示。

图 1.3　输入游戏者名字

另外，选择菜单项"排行榜"，可以显示各级别的最好成绩和创造该成绩纪录的游戏者名字，如图 1.4 所示。

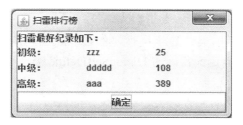

图 1.4　扫雷排行榜

7. 重新开始

在任何情况下，只要单击窗口上方中间的"重新开始"按钮，都可以重新开始扫雷游戏。

8. 加快扫雷进程

如果某个小方块周围的雷都已经标记出来（例如周围雷数是 3，在其相邻的 8 个小方块中已经有 3 个标记为雷），则在该小方块上单击鼠标右键，会将其相邻的没有标记为雷的小方块翻开。当然如果标记错误，在翻开的过程中就会踩到雷而导致扫雷失败。另外，在翻开小方块的过程中，如果某个小方块周围的雷数是 0，则也将其周围尚未翻开的小方块翻开，从而加快扫雷进程。

9. 附加功能

在扫雷的基本功能之外，还可以实现一些附加功能，如扫雷成功或失败时加入不同的声音；在一次单击翻开多个小方块时，也可以加入表示翻开的声音。

1.1.2　类的设计

根据以上的功能分析，设计出系统中需要的类。

1. MineFrame 类

MineFrame 类实现游戏的主窗口，包括菜单、窗口上方的面板（显示计时、雷数和重新开始的按钮），以及游戏的主要部分（雷区）。

2. MinePanel 类

MinePanel 类代表雷区，实现扫雷的主要功能，主要属性包括 Block 类型的二维对象数组。

3. Block 类

Block 类是雷区的小方块类，每个小方块有自己的类型、状态等属性，还有翻开、画出方块等方法。

4. BlockType 类和 BlockState 类

BlockType 类定义一组常量，表示小方块的类型（数值 0～8 分别表示小方块周围的雷数，数值 9 表示该小方块下面就是雷）。BlockState 类定义一组常量，表示小方块的状态（0～4 分别表示原始状态、翻开状态、标记为雷状态、标记为问号状态和雷爆炸状态）。

5. Record 类和 Grade 类

Record 类用于记录最好成绩，包括最好成绩创造者的名字和所用时间。Grade 类定义一组常量，表示游戏的难度级别（0、1、2、3 分别表示初级、中级、高级和自定义）。

6. RecordDao 类

RecordDao 类用于将 Record 对象输出到文件或从文件中读取 Record 对象。

7. 对话框类

扫雷游戏包括 3 个对话框类：显示排行榜的 DialogShowRecord 类、用户输入姓名的 DialogRecordName 类、用户自定义雷区的 DialogSelfDefine 类。

8. 监听器类

监听器类主要有监听菜单的监听器类、监听按钮的监听器类、监听鼠标的监听器类。

将扫雷游戏中几个主要类之间的关系用图 1.5 表示，其中 MineFrame、MinePanel、Block、Record 是本系统设计的主要类，JFrame、JPanel、Thread 是 JDK 提供的类，Serializable 是 JDK 提供的接口。

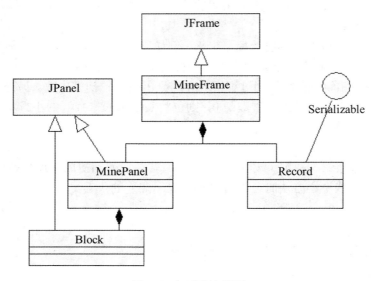

图 1.5 扫雷游戏类图

其他一些实现辅助功能的类并没有在图中绘出，在后面的系统实现中再详细介绍。

1.2 创建主窗口

扫雷游戏主窗口由菜单、窗口上方的计时区，以及中心部分的雷区组成，如图 1.6 所示。本节完成菜单和计时区的设计，雷区部分的设计将在 1.3 节实现。

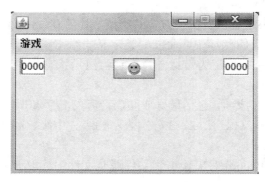

图 1.6 扫雷游戏主窗口

1.2.1 创建项目

启动 Eclipse 后，创建一个名为 Saolei 的 Java Project，然后在项目文件夹下创建子文件夹 image，将所需要的图标复制到该文件夹下，如图 1.7 所示。

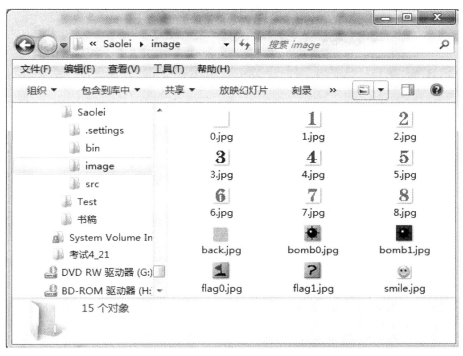

图 1.7 准备程序需要的图标

1.2.2 MineFrame 类

在 Saolei 项目中创建 MineFrame 类（从 JFrame 类继承），在类中添加菜单、按钮、文本框等组件，在构造方法中创建这些组件对象，并将这些组件组织到窗口中，代码如下：

```
/**
 * 扫雷游戏主窗口类
 */
public class MineFrame extends JFrame {
    JMenuBar menuBar;
    JMenu menu;
    JMenuItem[] menuItems;
    String[] menuItemNames = {"初级","中级","高级","自定义","排行榜","退出"};
    JTextField minesRemained;      //显示剩余雷数的文本框
    JButton reStart;               // "重新开始"按钮
    JTextField timeUsed;           //显示游戏使用时间的文本框
    Icon face;                     //按钮上的图标
    JPanel upPanel;                //计时区域
    public MineFrame(){
```

```java
        createMenu();
        createUpPanel();
        Container c =this.getContentPane();
        c.add(upPanel, BorderLayout.NORTH);
        this.setSize(300,200);
        this.setDefaultCloseOperation(EXIT_ON_CLOSE);
        this.setLocationRelativeTo(null);
        this.setResizable(false);
        this.setVisible(true);
    }
    /**
     * 创建计时区
     */
    private void createUpPanel(){
        minesRemained = new JTextField("0000");
        minesRemained.setEditable(false);
        timeUsed = new JTextField("0000");
        timeUsed.setEditable(false);
        face = new ImageIcon("image/smile.jpg");
        reStart = new JButton(face);
        reStart.addActionListener(new ButtonMonitor());
        JPanel center = new JPanel();
        JPanel right = new JPanel();
        JPanel left = new JPanel();
        center.add(reStart);
        left.add(minesRemained);
        right.add(timeUsed);
        upPanel = new JPanel(new BorderLayout());
        upPanel.add(left,BorderLayout.WEST);
        upPanel.add(center,BorderLayout.CENTER);
        upPanel.add(right,BorderLayout.EAST);
    }
    /**
     * 创建菜单并注册监听器
     */
    private void createMenu(){
        menuBar = new JMenuBar();
        menu = new JMenu("游戏");
        menuItems = new JMenuItem[menuItemNames.length];
        for(int i =0; i<menuItemNames.length; i++){
            menuItems[i] = new JMenuItem(menuItemNames[i]);
            menu.add(menuItems[i]);
        }
        menuBar.add(menu);
        this.setJMenuBar(menuBar);
        MenuMonitor mm = new MenuMonitor();
        for(int i=0; i<menuItems.length; i++){
```

```java
                menuItems[i].addActionListener(mm);
            }
        }
        /**
         * 按钮监听器类
         */
        class ButtonMonitor implements ActionListener{
            public void actionPerformed(ActionEvent e) {
            }
        }
        /**
         * 菜单监听器类
         */
        class MenuMonitor implements ActionListener{
            public void actionPerformed(ActionEvent e) {
                JMenuItem mi = (JMenuItem) e.getSource();
                if(mi.equals(menuItems[5])){
                    System.exit(0);
                }
            }
        }
    }
```

方法 createUpPanel()创建窗口上方的计时区 upPanel。首先创建两个文本框和一个按钮，然后将这 3 个组件分别放在各自的 JPanel 中，再将 3 个 Jpanel 组装到 upPanel 中。

方法 createMenu()创建菜单。Swing 的菜单系统由 JMenuBar、JMenu 和 JMenuItem 组成，扫雷游戏的菜单有 6 个菜单项（由数组 menuItemNames 确定）。首先创建这 6 个菜单项，并将菜单项添加到菜单 JMenu 中，然后将菜单添加到菜单条 JMenuBar 中，再调用 setJMenuBar()方法将菜单条添加到主窗口中，最后为所有的菜单项注册监听器。

内部类 ButtonMonitor 是按钮监听器，目前其 actionPerformed()方法没有代码，具体功能在后面实现。

内部类 MenuMonitor 是菜单监听器，目前只实现了"退出"菜单项的功能，其他菜单项的功能在后面实现。

在构造方法中调用 createMenu()方法和 createUpPanel()方法，创建菜单和计时区，设置窗口的关闭功能。方法 setLocationRelativeTo(null)的作用是设置窗口的相对位置，参数 null 表示相对于屏幕居中。

1.2.3 主程序类 Saolei

添加 Saolei 类，代码如下：

```java
public class Saolei {
    public static void main(String[] args) {
        MineFrame mineFrame = new MineFrame();
    }
}
```

在 main()方法中创建一个 MineFrame 对象，显示主窗口。运行这个程序的结果如图 1.6 所示。此时选择菜单项"退出"或者单击窗口右上角的关闭按钮，都能够退出程序。

1.3　MinePanel 类和 Block 类

MinePanel 类代表雷区，由于雷区是由若干行、若干列小方块组成的，因此要首先设计 Block 类。下面首先添加几个辅助类，然后再介绍 Block 类和 MinePanel 类。MinePanel 类设计好之后，扫雷的主窗口就全部实现了，运行后的界面（游戏难度初级，雷区大小是 10 行、10 列）如图 1.8 所示。

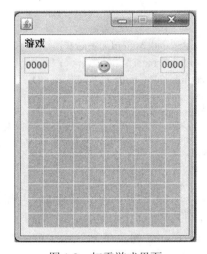

图 1.8　扫雷游戏界面

1.3.1　几个辅助类

1. BlockType 类

BlockType 类中定义一组常量，表示小方块的类型，数值 0~8 分别表示与该小方块相邻的雷数，数值 9 表示该小方块下面就是雷。由于每个小方块有 8 个相邻的小方块，因此相邻的雷数的可能值是 0~8，代码如下：

```
public class BlockType {
    static final int ZERO = 0;    //周围的雷数
    static final int ONE = 1;
    static final int TWO = 2;
    static final int THREE = 3;
    static final int FOUR = 4;
    static final int FIVE = 5;
    static final int SIX = 6;
    static final int SEVEN = 7;
    static final int EIHHT = 8;
    static final int MINE = 9;    //是雷
}
```

2. BlockState 类

BlockState 类中定义一组常量，表示小方块的状态，扫雷游戏中的小方块共有 5 种状态，分别是原始状态、翻开状态、标记为雷状态、标记为问号状态、雷爆炸状态，代码如下：

```
Public class blockstate {
    Static final int ORIGINAL = 0;           //初始状态
    Static final int OPEN = 1;               //翻开状态
    Static final int MINE_FLAG = 2;          //标记为雷状态
    Static final int QUESTION_FLAG = 3;      //标记为问号状态
    Static final int EXPLODED = 4;           //爆炸状态
}
```

3. Grade 类

Grade 类中定义一组常量，表示游戏的难度级别，扫雷游戏中的难度分为 4 个级别，分别是初级、中级、高级和自定义，代码如下：

```
public class Grade {
    public  static final int LOWER = 1;          //初级
    public  static final int MEDIAL = 2;         //中级
    public  static final int HIGHER = 3;         //高级
    public  static final int SELF_DEFINE = 4;    //自定义
}
```

BlockType 类和 BlockState 类是本节中用到的两个类，Grade 类在后面要用到。

1.3.2 Block 类

创建小方块类 Block，该类从 JPanel 类继承，由于小方块是不能单独存在的，一定是某个雷区中的小方块，因此在 Block 类中定义一个雷区类 MinePanel 类型的属性，其他属性包括该小方块在雷区中的行列号、小方块的宽和高、小方块的类型和状态等，代码如下：

```
public class Block extends JPanel {
    private MinePanel minePanel;
    private int row;                       //在雷区中的行号
    private int col;                       //在雷区中的列号
    public final int WIDTH = 19;           //方块的宽度
    public final int HEIGHT = 19;          //方块的高度
    private int type;                      //0,1,2,3,4,5,6,7,8,9（雷）
    private int state;                     //0 原始状态，1 翻开，2 标记为雷，3 标记为问号
    public static Toolkit tk;
    public static final Image[] numberImage;   //0~8
    public static final Image[] flagImage ;    //0 标记为雷，1 标记为问号
    public static final Image[] bombImage;     //0 未爆炸，1 已爆炸
    public static final Image backImage;       //未翻开时的背面
    static{ //静态代码段装载图标资源
        tk = Toolkit.getDefaultToolkit();
        numberImage = new Image[9];
        flagImage = new Image[2];
        bombImage = new Image[2];
        for(int i=0; i<numberImage.length; i++){
```

```java
            String fileName = "image/"+i+".jpg";
            numberImage[i] = tk.getImage(fileName);
        }
        for(int i=0; i<flagImage.length; i++){
            String fileName = "image/flag"+i+".jpg";
            flagImage[i] = tk.getImage(fileName);
        }
        for(int i=0; i<bombImage.length; i++){
            String fileName = "image/bomb"+i+".jpg";
            bombImage[i] = tk.getImage(fileName);
        }
        backImage = tk.getImage("Image/back.jpg");
    }
    public Block(){
        
    }
    public Block(MinePanel minePanel,int row, int col, int type, int state) {
        super();
        this.minePanel = minePanel;
        this.row = row;
        this.col = col;
        this.type = type;
        this.state = state;
    }
    public int getType() {
        return type;
    }
    public void setType(int type) {
        this.type = type;
    }
    public int getState() {
        return state;
    }
    public void setState(int state) {
        this.state = state;
    }
    /**
     * 翻开小方块
     * @return   true：翻开成功；false：翻开失败
     */
    public boolean open(){
        if(type!=BlockType.MINE){
            state = BlockState.OPEN;
            draw(minePanel.getGraphics());
            return true;
        }
```

```java
            else{
                state = BlockState.EXPLODED;
                draw(minePanel.getGraphics());
                return false;
            }
        }
    }
    /**
     * 显示小方块
     * @param g    来源于 MinePanel
     */
    public void draw(Graphics g){
        int x = col*minePanel.GRID_WIDTH;
        int y = row*minePanel.GRID_HEIGHT;
        switch(state){
            case BlockState.ORIGINAL:
                g.drawImage(backImage,x,y,WIDTH,HEIGHT,minePanel);
                break;
            case BlockState.MINE_FLAG:
                g.drawImage(flagImage[0],x, y, WIDTH,HEIGHT,minePanel);
                break;
            case BlockState.QUESTION_FLAG:
                g.drawImage(flagImage[1],x, y, WIDTH,HEIGHT,minePanel);
                break;
            case BlockState.OPEN:
                if(type==BlockType.MINE){
                    g.drawImage(bombImage[0],x, y, WIDTH,HEIGHT,minePanel);
                }
                else{
                    g.drawImage(numberImage[type],x, y, WIDTH,HEIGHT,minePanel);
                }
                break;
            case BlockState.EXPLODED:
                if(type==BlockType.MINE){
                    g.drawImage(bombImage[1],x, y, WIDTH,HEIGHT,minePanel);
                }
                break;
        }
    }
}
```

静态代码块用于装载图标资源，除了构造方法和一组 get、set 方法外，Block 只有 draw() 和 open()两个方法。

draw()方法是将小方块在雷区中显示出来，首先获取小方块的左上角坐标，然后根据不同的状态显示不同的图标。如果是原始状态，则显示小方块的背面；如果是标记为雷的状态，则显示小旗图标；如果是怀疑为雷的状态，则显示问号；如果是翻开状态，则显示具体数字图标；如果是爆炸状态，则显示爆炸图标。

open()方法是将小方块翻开，如果不是雷，将其状态设置为翻开状态并重新显示，方法返回 true；如果是雷，将其状态设置为爆炸状态并重新显示，方法返回 false。

1.3.3　MinePanel 类

MinePanel 类代表雷区，扫雷功能主要是在这个类中实现的。类中有两个表示小方块大小的常量（注意，这个常量值要比 Block 类中定义的值大一点，目的是在小方块之间留有一点空隙），一个 MineFrame 类型的属性，表示雷区行数、列数和雷数的 3 个属性，属性 remainedMines 表示尚未标记出来的雷数，属性 openedBlocks 表示已经翻开的小方块数，数组 blocks 保存所有的小方块，代码如下：

```java
public class MinePanel extends JPanel {
    public final int GRID_WIDTH = 20;      //方格的宽度
    public final int GRID_HEIGHT = 20;     //方格的高度
    private MineFrame mf;
    private int cols;              //雷区的列数
    private int rows;              //雷区的行数
    private int mines;             //雷区的雷数
    private int remainedMines;     //未标记的雷数
    private int openedBlocks;      //已经翻开的方块数
    private Block[][] blocks;      //方块数组
    public MinePanel(MineFrame mf, int rows, int cols, int mines) {
        this.mf = mf;
        initMinePanel(rows, cols, mines);
    }
    /**
     * 初始化雷区，创建雷区所有方块并随机布雷
     * @param rows     雷区行数
     * @param cols     雷区列数
     * @param mines    雷数
     */
    public void initMinePanel(int rows, int cols, int mines){
        this.cols = cols;
        this.rows = rows;
        this.mines = mines;
        remainedMines = mines;
        openedBlocks = 0;
        createBlocks();
        layMines();
        repaint();
    }
    /**
     * 创建 blocks 数组和数组中的对象，小方块类型初始化为 0，状态为初始状态 0
     */
    private void createBlocks(){
        blocks = new Block[rows][cols];
```

```java
            for(int i=0; i<rows; i++){
                for(int j=0; j<cols; j++){
                    blocks[i][j]= new Block(this,i,j,BlockType.ZERO,BlockState.ORIGINAL);
                }
            }
        }
        /**
         * 随机布雷,包括:初始化雷区、布雷、计算每个方块周围的雷数
         */
        private void layMines(){
            int r;
            int c;
            //初始化雷区
            for(int i=0; i<rows; i++){
                for(int j=0; j<cols; j++){
                    blocks[i][j].setType(BlockType.ZERO);
                    blocks[i][j].setState(BlockState.ORIGINAL);
                }
            }
            //随机布雷
            int m = 0;
            while(m<mines){
                r = (int) (Math.random()*rows);
                c = (int) (Math.random()*cols);
                if(blocks[r][c].getType()!=BlockType.MINE){
                    blocks[r][c].setType(BlockType.MINE);
                    m++;
                }
            }
            //计算每个 block 周围的雷数
            for(int i=0; i<rows; i++){
                for(int j=0; j<cols; j++){
                    if(blocks[i][j].getType()!=BlockType.MINE){
                        countMines(i,j);
                    }
                }
            }
        }
        /**
         * 计算小方块周围的雷数
         * @param row    小方块的行号
         * @param col    小方块的列号
         */
        private void countMines(int row, int col){
            int mineNumber=0;   //blocks[i][j]周围的雷数
```

```java
            for(int i=row-1; i<=row+1;i++){   //在 block 的周围搜索(上下两行)
                if( (i>=0)&&(i<rows) ){   //判断没有越界(数组下标越界)
                    for(int j=col-1; j<=col+1; j++){   //搜索左右两列
                        if( (j>=0)&&(j<cols) ){    //判断没有越界
                            if(blocks[i][j].getType()==BlockType.MINE){
                                mineNumber++;
                            }
                        }
                    }
                }
            }
            blocks[row][col].setType(mineNumber);
    }
    /**
     * 重写 paint 方法,显示所有方块
     */
    public void paint(Graphics g) {
        super.paint(g);
        for(int i=0; i<rows; i++){
            for(int j=0; j<cols; j++){
                blocks[i][j].draw(g);
            }
        }
    }
    /**
     * 重写 getPreferredSize 方法
     */
    public Dimension getPreferredSize() {
        return new Dimension(cols * GRID_WIDTH, rows*GRID_HEIGHT);
    }
}
```

在构造方法中调用 initMinePanel()方法对雷区初始化。

initMinePanel()方法有 3 个参数,用来初始化雷区的行数、列数和雷数。在 initMinePanel() 方法中将 remainedMines 设置为雷数,将 openedBlocks 设置为 0,然后调用 createBlocks()方法创建雷区的所有小方块,调用 layMines()方法随机布雷。

在随机布雷 layMines()方法中,首先初始化雷区,然后随机布雷,最后调用 countMines(i,j) 方法计算小方块周围的雷数。

paint()方法将雷区中的小方块显示出来,getPreferredSize()方法计算雷区最适合的大小,后面的程序将在 MineFrame 中让主窗口的大小自动适应其内部组件的大小时自动调用这个方法。

1.3.4 将雷区加入到 MineFrame 中

为了把雷区加入到主窗口中,需要在 MineFrame 类中添加 MinePanel 类型的属性,在构造

方法中创建 MinePanel 对象,并将其添加到主窗口中。由于在创建 MinePanel 对象时要指定雷区的行列数和雷数,我们在 MineFrame 类中添加 3 个整型属性,并添加一个方法 initParameter() 为行列数和雷数赋值。

1. 在 MineFrame 类中添加属性

在 MineFrame 类中添加以下 4 个属性:

```
MinePanel minePanel;
private int rows;
private int cols;
private int mines;
```

2. 添加 initParameter 方法

在 MineFrame 类中添加 initParameter()方法,代码如下:

```
private void initParameter(int rows, int cols, int mines){
    this.rows = rows;
    this.cols = cols;
    this.mines = mines;
}
```

3. 修改构造方法

修改后的构造方法如下:

```
public MineFrame(){
    createMenu();
    createUpPanel();
    initParameter(10,10,10);
    Container c =this.getContentPane();
    c.add(upPanel, BorderLayout.NORTH);
    centerPanel = new MinePanel(this, rows,cols,mines);
    JPanel centerPanel = new JPanel();
    centerPanel.setLayout(new BorderLayout());
    centerPanel.add(new JPanel(),BorderLayout.WEST);
    centerPanel.add(minePanel,BorderLayout.CENTER);
    minePanel.add(new JPanel(),BorderLayout.EAST);
    c.add(centerPanel,BorderLayout.CENTER);
    this.pack();
    this.setDefaultCloseOperation(EXIT_ON_CLOSE);
    this.setLocationRelativeTo(null);
    this.setResizable(false);
    this.setVisible(true);
}
```

修改的代码包括调用 initParameter()方法为行列数和雷数初始化,默认难度级别为初级,3 个参数都是 10;创建 MinePanel 对象,并将其组装到主窗口中;调用 pack()方法使窗口的大小自动适应其内部组件的大小。

在将 minePanel 组装到主窗口时,首先创建 centerPanel 对象,将 3 个 Panel 对象按左中右的结构添加到 centerPanel 中,再将 centerPanel 添加到主窗口中。这样做的目的是让系统自动处理 BorderLayout 布局中左右两部分的大小。

1.4 实现扫雷功能

1.4.1 翻开小方块

单击鼠标左键翻开小方块，如果是雷则游戏失败，否则翻开。

1. 在 MineFrame 类中添加属性

在 MineFrame 类中添加表示正在扫雷和游戏停止的两个属性，代码如下：

```
private boolean gamming;    //正在扫雷
private boolean stoped;     //游戏已停止
```

2. 在 MineFrame 类中添加 set、get 方法

在 MineFrame 类中添加一组 set、get 方法，代码如下：

```
public boolean isGamming() {
    return gamming;
}
public void setGamming(boolean gamming) {
    this.gamming = gamming;
}
public boolean isStoped() {
    return stoped;
}
public void setStoped(boolean stoped) {
    this.stoped = stoped;
}
public int getRows() {
    return rows;
}
public int getCols() {
    return cols;
}
public int getMines() {
    return mines;
}
public void setMinesRemained(int mines){
    String strMines;
    if(mines>9999){
        strMines = "9999";
    }
    else if(mines/10 == 0){
        strMines = "000"+mines;
    }
    else if (mines/100 == 0){
        strMines = "00"+mines;
```

```
            }
            else if (mines/1000 == 0){
                strMines = "0"+mines;
            }
            else{
                strMines = ""+mines;
            }
            minesRemained.setText(strMines);
        }
        public void setTimeUsed(int second){
            String strSecond;
            if(second>9999){
                strSecond = "9999";
            }
            else if(second/10 == 0){
                strSecond = "000"+second;
            }
            else if (second/100 == 0){
                strSecond = "00"+second;
            }
            else if (second/1000 == 0){
                strSecond = "0"+second;
            }
            else{
                strSecond = ""+second;
            }
            timeUsed.setText(strSecond);
        }
```

方法 setMinesRemained()是设置剩余雷数文本框的数值，文本框一共显示 4 位字符，如果参数为 1 位数字，则在前面加 3 个 "0"；如果参数为 2 位数字，则在前面加 2 个 "0"；如果参数为 3 位数字，则在前面加 1 个 "0"；如果参数为 4 位数字，则不需要加 "0"。

方法 setTimeUsed()是设置游戏经历时间文本框的数值，与方法 setMinesRemained()类似。

3. 修改 MineFrame 类的 initParameter()方法

修改 MineFrame 类的 initParameter()方法，给 stoped 和 gamming 赋值为 false，并设置两个文本框的初始值，修改后的代码如下：

```
        private void initParameter(int rows, int cols, int mines){
            this.rows = rows;
            this.cols = cols;
            this.mines = mines;
            stoped = false;
            gamming = false;
            setTimeUsed(0);
            setMinesRemained(mines);
        }
```

4. 在 MinePanel 类中添加鼠标监听器类

在 MinePanel 类中添加内部监听器类，用来监听鼠标消息，代码如下：

```java
/**
 * 监听鼠标消息
 */
class MouseMonitor extends MouseAdapter{
    public void mouseClicked(MouseEvent event) {
        int col = event.getX()/GRID_WIDTH;
        int row = event.getY()/GRID_HEIGHT;
        if(mf.isStoped())        //如果游戏已停止
            return;
        //如果原来不是游戏状态，则开始游戏，设置游戏进行中
        if(!mf.isGamming()){
            mf.setGamming(true);
        }
        //如果是左键，翻开小方块
        if(event.getButton() == MouseEvent.BUTTON1){
            open(row,col);
        }
        else if(event.getButton() == MouseEvent.BUTTON3){
            if(blocks[row][col].getState()==BlockState.ORIGINAL){
                blocks[row][col].setState(BlockState.MINE_FLAG);
                remainedMines--;
                mf.setMinesRemained(remainedMines);
                blocks[row][col].draw(MinePanel.this.getGraphics());
            }
            else if(blocks[row][col].getState()==BlockState.MINE_FLAG){
                blocks[row][col].setState(BlockState.QUESTION_FLAG);
                remainedMines++;
                mf.setMinesRemained(remainedMines);
                blocks[row][col].draw(MinePanel.this.getGraphics());
            }
            else if(blocks[row][col].getState()==BlockState.QUESTION_FLAG){
                blocks[row][col].setState(BlockState.ORIGINAL);
                blocks[row][col].draw(MinePanel.this.getGraphics());
            }
        }
    }
}
```

如果是鼠标左键，则调用 open()方法翻开小方块（open()方法稍后介绍）；如果是鼠标右键，则修改小方块的标记。如果小方块是原始状态，则将其标记为雷，未标记的雷数减 1，更新剩余雷数文本框的数值，重新显示该小方块；如果小方块是标记为雷的状态，则将其标记为问号，未标记的雷数加 1，更新剩余雷数文本框的数值，重新显示该小方块；如果小方块标记为问号，则将其恢复为原始状态，重新显示该小方块。

5. 在 MinePanel 类中添加 open()方法

在 MinePanel 类中添加翻开小方块的方法 open()，代码如下：

```
/**
 * 翻开指定的小方块
 * @param row    小方块的行号
 * @param col    小方块的列号
 */
public void open(int row, int col){
    //只有原始状态才处理，如果已经翻开或已标记则忽略
    if(blocks[row][col].getState()==BlockState.ORIGINAL){
        if(blocks[row][col].open()){
            openedBlocks++;
            if(openedBlocks == rows * cols -   mines){
                //wins();
            }
            if(blocks[row][col].getType()==BlockType.ZERO){
                //search(row, col);   //在周围搜索不是雷的方块
            }
        }else{
            //lose(row,col);       //失败处理
        }
    }
}
```

如果小方块是原始状态，则调用 Block 类的 open()方法翻开小方块。如果成功，则将翻开小方块的数目加 1，如果翻开的小方块数目是小方块总数减去雷数，则小方块已经全部翻开，调用 wins()方法完成扫雷成功后的处理（wins()方法稍后给出，这里暂时将这行代码注释掉）。如果翻开的小方块周围没有雷，则调用 search()方法搜索并翻开周围的小方块，尽可能翻开更多的小方块（search()方法稍后给出，这里暂时将这行代码注释掉）。

如果翻开小方块失败，则表明踩中雷，扫雷失败，调用 lose()方法完成扫雷失败后的处理（lose()方法也在稍后给出，这里暂时将这行代码注释掉）。

6. 注册监听器

在 MinePanel 类的构造方法中注册监听器并设置背景颜色，使小方块翻开后小方块之间的间隙颜色与小方块翻过来的颜色不一致。修改后的构造方法如下：

```
public MinePanel(MineFrame mf, int rows, int cols, int mines) {
    this.mf = mf;
    initMinePanel(rows, cols, mines);
    this.addMouseListener(new MouseMonitor());
    this.setBackground(new Color(210,210,210));
}
```

现在重新运行程序，单击鼠标左键可以翻开小方块，单击鼠标右键可以为小方块添加标记，如图 1.9 所示。

图 1.9　小方块的翻开与标记

1.4.2　处理输赢以及搜索方法

为实现扫雷的完整功能，在 MinePanel 类中添加扫雷成功或失败的处理方法 wins()和 lose()，以及加快扫雷速度的 search()方法。

1. wins()方法

wins()方法完成扫雷成功后的处理，代码如下：

```
/**
 * 扫雷结束的处理
 */
public void wins() {
    mf.setGamming(false);
    mf.setStoped(true);
    JOptionPane.showMessageDialog(this, "恭喜，扫雷成功！");
}
```

扫雷成功后，设置正在游戏属性为 false，停止游戏属性为 true，并显示扫雷成功的信息。

2. lose()方法

lose()方法实现扫雷失败后的处理，代码如下：

```
/**
 * 踩到雷后的处理
 * @param row
 * @param col
 */
public void lose(int row, int col) {
    int i,j;
    //将没有标记为雷的雷翻开
    for(i=0; i<rows; i++){
        for(j=0; j<cols; j++){
            if( (blocks[i][j].getType()==BlockType.MINE) &&
                (blocks[i][j].getState()!=BlockState.MINE_FLAG)){
```

```
                    blocks[i][j].setState(BlockState.OPEN);
                    blocks[i][j].draw(MinePanel.this.getGraphics());
                }
            }
        }
        blocks[row][col].setState(BlockState.EXPLODED);
        blocks[row][col].draw(MinePanel.this.getGraphics());
        mf.setGamming(false);
        mf.setStoped(true);
        JOptionPane.showMessageDialog(this, "扫雷失败，继续努力！");
    }
```

扫雷失败后，将是雷但没有标记出来的小方块翻开，将踩中雷的小方块设置为爆炸状态，然后设置正在游戏属性为 false，停止游戏属性为 true，并显示扫雷失败的信息。

3. search()方法

search()是一个递归方法，功能是尽可能多地翻开小方块，代码如下：

```
/**
 * 将周围雷数为 0 的小方块的相邻小方块都翻开，在翻开的过程中
 * 如果再遇到周围雷数为 0 的小方块，继续将其相邻的小方块翻开
 * @param row    小方块的行号
 * @param col    小方块的列号
 */
private void search(int row, int col){
    int i,j;
    //在周围搜索
    for(i=row-1; i<=row+1; i++){
        if( (i<0) || (i>=rows) ){  //行号超范围
            continue;
        }
        for(j=col-1; j<=col+1; j++){
            if( (j<0)||(j>=cols) ){ //列号超范围
                continue;
            }
            if( blocks[i][j].getState()==BlockState.ORIGINAL ){
                //在States.ZERO 周围肯定没有雷
                blocks[i][j].open();
                openedBlocks++;
                if(openedBlocks == rows * cols - mines){
                    wins();
                }
                if(blocks[i][j].getType()==BlockType.ZERO){
                    search(i,j);   //递归搜索
                }
            }
        }
    }
}
```

search()方法将周围雷数为 0 的小方块的相邻小方块翻开,使用递归方法,在翻开的过程中,遇到周围雷数为 0 的小方块,继续搜索翻开。

4. 去掉 open()方法中的注释

将 1.4.2 节 open()方法中的注释去掉,即将调用 wins()方法、lose()方法和 search()方法的 3 个语句的注释去掉。

运行程序,单击小方块,如果单击的是周围雷数为 0 的小方块,其周围的小方块也会自动翻开,并且在成功或失败时会显示相关的信息。

1.4.3 加快扫雷进程

在扫雷过程中,某个小方块周围的雷都已标记出来,此时该小方块的相邻小方块如果还有未标记未翻开的小方块,就可以直接翻开了。例如图 1.10 中两个标记为雷的小方块中间显示为"2"的小方块,由于该小方块周围的两个雷都已标记出来,其上方的三个小方块肯定不是雷,可以通过右击该小方块将其上方的三个小方块翻开。

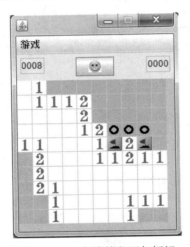

图 1.10 小方块的翻开与标记

在前面的程序中,已经添加了 MinePanel 的内部监听器类,该监听器已经实现了左键翻开小方块、右键标记小方块。如果右键单击了尚未翻开的小方块,则标记小方块。下面增加一个功能,如果右击的是已经翻开的小方块,则翻开其相邻的没有标记为雷的小方块。修改鼠标监听器类 MouseMonitor,在右键单击的分支中再增加一个小方块为翻开状态的分支,增加以下的代码(最上面的 4 行代码是前面就有的):

```
        else  if(blocks[row][col].getState()==BlockState.QUESTION_FLAG){
            blocks[row][col].setState(BlockState.ORIGINAL);
            blocks[row][col].draw(MinePanel.this.getGraphics());
        }
        else if(blocks[row][col].getState()==BlockState.OPEN){
            //如果小方块是已经翻开的且其周围的雷都已标记出来,则将与其相邻没有标记的小方块翻开
            int flagNumber = 0;
            for(int i=row-1; i<=row+1; i++){    //计算已经标记为雷的相邻方块数
                for(int j=col-1; j<=col+1; j++){
```

```
                if((i>=0)&&(i<rows)&&(j>=0)&&(j<cols)){
                    if(blocks[i][j].getState() == BlockState.MINE_FLAG){
                        flagNumber++;
                    }
                }
            }
        }
        if(flagNumber == blocks[row][col].getType()){//相邻的雷都已经标记出来
            for(int i=row-1; i<=row+1; i++){
                for(int j=col-1; j<=col+1; j++){
                    if((i>=0)&&(i<rows)&&(j>=0)&&(j<cols)){
                        open(i,j);
                    }
                }
            }
        }
    }
```

上面新增加的代码中，第一个循环计算周围已经标记为雷的小方块数，第二个循环调用 open()方法，翻开其相邻的小方块。如果前面对雷的标记有误，在翻开小方块时就会踩到雷导致扫雷失败。

1.4.4 重新开始游戏

前面已经在 MineFrame 类中添加了内部按钮监听器类，现在只要在其 actionPerformed() 方法中写入代码即可，代码如下：

```
    /**
     * 按钮监听器类
     */
    class ButtonMonitor implements ActionListener{
        public void actionPerformed(ActionEvent e) {
            initParameter(rows, cols, mines);
            minePanel.initMinePanel(rows, cols, mines);
            minePanel.repaint();
        }
    }
```

actionPerformed()方法中调用了初始化参数的两个方法，对 MineFrame 类和 MinePanel 类中的游戏参数进行初始化，然后更新显示，将 stoped 和 gamming 都设置为 false。

至此扫雷的基本功能都已实现，下一节将实现通过菜单选择游戏难度级别的功能。

1.5 选择游戏难度级别

1.5.1 在 MineFrame 类中添加 grade 属性

1. 添加 grade 属性

在 MineFrame 类中添加 grade 属性，表示游戏的难度级别，代码如下：

```
        private int grade;
```

2. 添加 get 和 set 方法

添加 grade 属性对应的 get、set 方法，代码如下：

```
    public int getGrade() {
        return grade;
    }
    public void setGrade(int grade) {
        this.grade = grade;
    }
```

1.5.2 自定义难度对话框

除了初级、中级、高级 3 个难度级别外，用户还可以通过菜单打开自定义雷区对话框来设置雷区的行列数和雷数。自定义雷区对话框如图 1.11 所示。

图 1.11 自定义雷区对话框

在对话框中输入行数、列数和雷数，单击"确定"按钮，将重新设置雷区的大小并布雷。
自定义雷区对话框类如下：

```java
public class DialogSelfDefine extends JDialog implements ActionListener{
    MineFrame mf;
    JLabel labelRows = new JLabel("行数：");
    JLabel labelCols = new JLabel("列数：");
    JLabel labelMines = new JLabel("雷数：");
    JTextField  tfRows = new JTextField(5);
    JTextField  tfCols = new JTextField(5);
    JTextField  tfMines = new JTextField(5);
    JButton btOK = new JButton("确定");
    JButton btCancel = new JButton("取消");
    private int option=0;
    public DialogSelfDefine(MineFrame parent){
        super(parent, "用户自定义",JDialog.ModalityType.APPLICATION_MODAL);
        mf = parent;
        this.setLayout(new BorderLayout());
        JPanel jpWest = new JPanel();
        JPanel jpCenter = new JPanel();
        JPanel jpSouth = new JPanel();
        jpWest.setLayout(new GridLayout(3, 1));
        jpCenter.setLayout(new GridLayout(3, 1));
```

```
                jpSouth.setLayout(new FlowLayout());
                jpWest.add(labelRows);
                jpWest.add(labelCols);
                jpWest.add(labelMines);
                jpCenter.add(tfRows);
                jpCenter.add(tfCols);
                jpCenter.add(tfMines);
                jpSouth.add(btOK);
                jpSouth.add(btCancel);
                tfRows.setText("" + mf.getRows());
                tfCols.setText("" + mf.getCols());
                tfMines.setText("" + mf.getMines());
                this.add(new JPanel(),BorderLayout.NORTH);       //上方留点空白
                this.add(jpWest,BorderLayout.WEST);
                this.add(jpCenter,BorderLayout.CENTER);
                this.add(new JPanel(),BorderLayout.EAST);        //右侧留点空白
                this.add(jpSouth,BorderLayout.SOUTH);
                btOK.addActionListener(this);
                btCancel.addActionListener(this);
                this.setDefaultCloseOperation(DISPOSE_ON_CLOSE);
                this.pack();
                this.setLocationRelativeTo(mf);
            }
            public void actionPerformed(ActionEvent e) {
                if(e.getSource().equals(btOK)){
                    option = 1;
                    dispose();
                }
                else if(e.getSource().equals(btCancel)){
                    option = 0;
                    dispose();
                }
            }
            public int openDialog() {
                this.setVisible(true);
                return option;
            }
        }
```

在 DialogSelfDefine 类的构造方法中，调用父类的构造方法设置对话框的父窗口和对话框的标题，并设置对话框为模式对话框。构造方法中的大部分代码用于创建对话框中的组件并组装到一起，最后为两个按钮注册监听器。

actionPerformed()方法处理鼠标单击事件，如果单击的是"确定"按钮，将 option 设置为 1；如果单击的是"取消"按钮，将 option 设置为 0。

openDialog()方法用于显示对话框并在对话框消失后返回 option 的值，根据返回值确定单击的是"确定"按钮还是"取消"按钮。

1.5.3 完善菜单监听器类

前面已经添加了菜单的监听器类，但只处理了"退出"菜单项。下面增加处理"初级""中级""高级""自定义"菜单项的处理分支，修改后的菜单监听器代码如下：

```java
class MenuMonitor implements ActionListener{
    public void actionPerformed(ActionEvent e) {
        JMenuItem mi = (JMenuItem) e.getSource();
        if(mi.equals(menuItems[5])){    //退出
            System.exit(0);
        }
        else if(mi.equals(menuItems[4])){ //排行榜
        }
        else {    //选择难度等级
            if(mi.equals(menuItems[0])){
                grade = Grade.LOWER;
                initParameter(10,10,10);
            }
            else if(mi.equals(menuItems[1])){
                grade = Grade.MEDIAL;
                initParameter(16,16,40);
            }
            else if(mi.equals(menuItems[2])){
                grade = Grade.HIGHER;
                initParameter(16,30,99);
            }
            else if(mi.equals(menuItems[3])){
                DialogSelfDefine dlg = new DialogSelfDefine(MineFrame.this);
                if(dlg.openDialog()==1){
                    grade = Grade.SELF_DEFINE;
                    int r = Integer.valueOf(dlg.tfRows.getText());
                    int c = Integer.valueOf(dlg.tfCols.getText());
                    int m = Integer.valueOf(dlg.tfMines.getText());
                    initParameter(r,c,m);
                }
            }
            minePanel.initMinePanel(rows, cols, mines);
            pack();
            MineFrame.this.setLocationRelativeTo(null);
        }
    }
}
```

有关排行榜的分支稍后处理。如果选择"初级""中级"或"高级"菜单项，直接设置游戏参数；如果选择"自定义"菜单项，则打开自定义对话框，输入各参数，如果返回值是 1，则单击的是"确定"按钮，根据文本框中的值初始化游戏参数，最后初始化 MinePanel 类中的参数。

1.6 实现计时功能

我们使用 JDK 提供的 Timer 类实现计时功能，通过 Timer 计时器可以设置指定时间间隔执行某项任务，这项任务可由 TimerTask 指定，为此首先创建我们自己的 TimeTask 类，也就是 UpdateTimeTask 类，然后在游戏开始时启动计时，在游戏结束时终止计时。

1.6.1 UpdateTimeTask 类

1. 创建 UpdateTimeTask 类

创建 UpdateTimeTask 类，该类继承于 TimerTask 类，用于游戏界面已用时间的更新，代码如下：

```java
public class UpdateTimeTask extends TimerTask {
    private MineFrame mf;
    private int second;
    public UpdateTimeTask(MineFrame mf) {
        super();
        this.mf = mf;
        this.second = 0;
    }
    public int getSecond() {
        return second;
    }
    public void run() {
        second++;
        mf.setTimeUsed(second);
    }
}
```

由于要更新主窗口的计时组件，因此类中包含一个 MineFrame 类型的属性，在构造方法中为其赋值。另一个属性 second 用于记录时间，将其初始化为 0。

方法 run() 每执行一次，将 second 的值加 1，并将时间值显示在窗口的计时文本框中。

2. 添加 UpdateTimeTask 类型的属性

在 MinePanel 类中增加 UpdateTimeTask 类型的属性如下：

```java
UpdateTimeTask utt;
```

1.6.2 启动计时与终止计时

1. 启动计时

在 MinePanel 类的菜单监听器类 MouseMonitor 中，增加启动计时器的代码，找到"if(!mf.isGamming()){"的位置，增加如下三行代码：

```java
if(!mf.isGamming()){
    mf.setGamming(true);
    utt = new UpdateTimeTask(mf);
    Timer timer = new Timer();
    timer.schedule(utt, 1000, 1000);
}
```

首先创建 UpdateTimeTask 对象和 Timer 对象，然后用 Timer 对象调用 schedule()方法，该方法的 3 个参数分别是要执行的任务、开始延迟时间、多长时间重复一次任务（这里我们指定为 1000 毫秒，即每隔一秒更新一次时间）。

2. 终止计时

在扫雷成功、失败、重新开始、菜单改变雷区时都要终止计时。
在 MinePanel 类的 wins()方法和 lose()方法的开始处添加下面这行代码来终止计时。
　　　　utt.cancel();
找到 MineFrame 类的内部监听器类 ButtonMonitor，在 actionPerformed()方法中的最前面添加下面这行代码。
　　　　if(minePanel.utt!=null)
　　　　　　minePanel.utt.cancel();
同样在 MenuMonitor 类的 actionPerformed()方法中找到选择游戏难度级别的代码分支开始处，添加下面这两行代码。
　　　　if(minePanel.utt!=null)
　　　　　　minePanel.utt.cancel();
调用 MinePanel 类中属性 utt 的 cancel()方法结束计时。至此，计时功能处理完毕。

1.7 扫雷排行榜

这里所说的排行榜指的是每个级别（初级、中级、高级）的最好成绩。为了实现扫雷排行榜的功能，需要创建一个游戏纪录类 Record（包括每个级别的游戏者名字、完成扫雷任务用时等属性）、一个处理 Record 对象的 RecordDao 类（将 Record 对象写入文件和从文件中读出 Record 对象）、一个用于输入游戏者名字的对话框类和一个显示排行榜的对话框类。

1.7.1 Record 类

Record 类保存排行榜的数据，代码如下：

```
public class Record implements Serializable {
    private static final long serialVersionUID = 9163774801769495403L;
    private String higherName;      //高级难度游戏者名字
    private int higherScore;        //高级难度纪录的用时
    private String medialName;      //中级难度游戏者名字
    private int medialScore;        //中级难度纪录的用时
    private String lowerName;       //低级难度游戏者名字
    private int lowerScore;         //低级难度纪录的用时
    public Record(){
    }
    public Record(String higherName, int higherScore, String medialName,
            int medialScore, String lowerName, int lowerScore) {
        this.higherName = higherName;
        this.higherScore = higherScore;
        this.medialName = medialName;
        this.medialScore = medialScore;
```

```java
            this.lowerName = lowerName;
            this.lowerScore = lowerScore;
        }
        public String getHigherName() {
            return higherName;
        }
        public void setHigherName(String higherName) {
            this.higherName = higherName;
        }
        public int getHigherScore() {
            return higherScore;
        }
        public void setHigherScore(int higherScore) {
            this.higherScore = higherScore;
        }
        public String getMedialName() {
            return medialName;
        }
        public void setMedialName(String medialName) {
            this.medialName = medialName;
        }
        public int getMedialScore() {
            return medialScore;
        }
        public void setMedialScore(int medialScore) {
            this.medialScore = medialScore;
        }
        public String getLowerName() {
            return lowerName;
        }
        public void setLowerName(String lowerName) {
            this.lowerName = lowerName;
        }
        public int getLowerScore() {
            return lowerScore;
        }
        public void setLowerScore(int lowerScore) {
            this.lowerScore = lowerScore;
        }
        public static long getSerialversionuid() {
            return serialVersionUID;
        }
    }
```

Record 类比较简单,除了一组属性外,只有构造方法和一组 get、set 方法。由于要将 Record 对象写入文件,因此 Record 类实现了 Serializable 接口。

1.7.2 RecordDao 类

RecordDao 类有两个方法，分别是将一个 Record 对象写入文件和从文件中读取一个 Record 对象，代码如下：

```java
public class RecordDao {
    private String fileName;
    public RecordDao(){
        fileName = "record/record.dat";
    }
    /**
     * 从文件中读取一个 Record 对象
     * @return  Record 对象
     */
    public Record readRecord(){
        FileInputStream    fis=null;
        ObjectInputStream ois=null;
        Record record = null;
        try {
            fis = new FileInputStream(fileName);
        } catch (FileNotFoundException e) {
            return null;
        }
        try {
            ois = new ObjectInputStream(fis);
            record = (Record) ois.readObject();
        } catch (IOException e) {
            e.printStackTrace();
        }catch (ClassNotFoundException e) {
            e.printStackTrace();
        }finally{
            try {
                ois.close();
            } catch (IOException e) {
                e.printStackTrace();
            }
            try {
                fis.close();
            } catch (IOException e) {
                e.printStackTrace();
            }
        }
        return record;
    }
    /**
     * 向文件输出一个 Record 对象
     */
```

```java
public void writeRecord(Record record){
    FileOutputStream  fos=null;
    ObjectOutputStream oos=null;
    try {
        fos = new FileOutputStream(fileName);
        oos = new ObjectOutputStream(fos);
        oos.writeObject(record);
    } catch (FileNotFoundException e) {
        e.printStackTrace();
    }catch(IOException e){
        e.printStackTrace();
    }
    finally{
        try {
            oos.close();
        } catch (IOException e) {
            e.printStackTrace();
        }
        try {
            oos.close();
        } catch (IOException e) {
            e.printStackTrace();
        }
    }
}
```

RecordDao 类有两个主要方法：readRecord()方法从文件中读取一个 Record 对象，writeRecord()方法写一个 Record 对象到文件中。我们指定的文件是 record/record.dat，因此要在项目所在的文件夹中创建一个子文件夹 record。如果这个文件夹不存在，则会产生异常，这里的两个方法并没有很好地处理这个异常，我们把这个问题当作一道作业题，请读者课后自行完成。

1.7.3 用于输入游戏者名字的对话框类

扫雷成功后，如果扫雷时间破纪录，则出现一个对话框，用于输入游戏者的名字。创建 DialogRecordName 类，代码如下：

```java
public class DialogRecordName extends JDialog implements ActionListener{
    private JLabel msg;
    JTextField name;
    private JButton OK;
    private JButton cancel;
    private int option=0;
    public DialogRecordName(MineFrame parent, int second,int grade){
        super(parent, "",JDialog.ModalityType.APPLICATION_MODAL);
        switch(grade){
            case Grade.LOWER:
                setTitle("初级成绩纪录");
```

```java
            break;
        case Grade.MEDIAL:
            setTitle("中级成绩纪录");
            break;
        case Grade.HIGHER:
            setTitle("高级成绩纪录");
            break;
    }
    msg= new JLabel("您的成绩是" + second + "秒, 破纪录, 请输入您的名字。");
    name =   new JTextField(20);
    OK = new JButton("确定");
    cancel = new JButton("取消");
    OK.addActionListener(this);
    cancel.addActionListener(this);
    this.setLayout(new GridLayout(2,1));
    this.add(msg);
    JPanel downPanel = new JPanel();
    downPanel.add(name);
    downPanel.add(OK);
    downPanel.add(cancel);
    this.add(downPanel);
    this.pack();
    this.setLocationRelativeTo(parent);
}
public void actionPerformed(ActionEvent e) {
    if(e.getSource().equals(OK)){
        option=1;
        dispose();
    }
    else{
        option=0;
        dispose();
    }
}
public int openDialog() {
    this.setVisible(true);
    return option;
}
}
```

openDialog()方法用于打开对话框并返回 option 的值。如果单击"确定"按钮，关闭对话框，监听器的 actionPerformed()方法将 option 设置为 1，否则设置为 0。因此可以根据其返回值来判断是单击的哪个按钮关闭的对话框。

1.7.4 显示排行榜的对话框类

创建显示排行榜的对话框类 DialogShowRecord，代码如下：

```java
public class DialogShowRecord extends JDialog implements ActionListener {
    MineFrame mf;
    JLabel title = new JLabel("");
    JLabel gradeLower = new JLabel("初级：");
    JLabel gradeMedial = new JLabel("中级：");
    JLabel gradeHigher = new JLabel("高级：");
    JLabel nameLower = new JLabel("");
    JLabel nameMedial = new JLabel("");
    JLabel nameHigher = new JLabel("");
    JLabel scoreLower = new JLabel("");
    JLabel scoreMedial = new JLabel("");
    JLabel scoreHigher = new JLabel("");
    JButton btOK = new JButton("确定");
    private Record record;
    public DialogShowRecord(MineFrame parent){
        super(parent, "扫雷排行榜",JDialog.ModalityType.APPLICATION_MODAL);
        this.mf = parent;
        this.setLayout(new BorderLayout());
        JPanel jpCentert = new JPanel();
        JPanel jpSouth = new JPanel();
        JPanel jpNorth = new JPanel();
        jpNorth.add(title);
        jpCentert.setLayout(new GridLayout(3,3));
        jpCentert.add(gradeLower);
        jpCentert.add(nameLower);
        jpCentert.add(scoreLower);
        jpCentert.add(gradeMedial);
        jpCentert.add(nameMedial);
        jpCentert.add(scoreMedial);
        jpCentert.add(gradeHigher);
        jpCentert.add(nameHigher);
        jpCentert.add(scoreHigher);
        jpSouth.add(btOK);
        this.add(title,BorderLayout.NORTH);
        this.add(jpCentert,BorderLayout.CENTER);
        this.add(btOK,BorderLayout.SOUTH);
        btOK.addActionListener(this);
        Record record = mf.getRecord();
        if(record!=null){
            title.setText("扫雷最好纪录如下：");
            nameLower.setText(record.getLowerName());
            nameMedial.setText(record.getMedialName());
            nameHigher.setText(record.getHigherName());
            scoreLower.setText("" + record.getLowerScore());
            scoreMedial.setText("" + record.getMedialScore());
            scoreHigher.setText("" + record.getHigherScore());
```

```
            }
            else{
                title.setText("未找到扫雷纪录");
            }
            this.setSize(300, 150);
            this.setLocationRelativeTo(parent);
            this.setVisible(true);
        }
        public void actionPerformed(ActionEvent arg0) {
            dispose();
        }
    }
```

这个对话框类也比较简单，就是将 record 记录的数据显示出来。

1.7.5 实现排行榜功能

程序运行时，将排行榜数据从文件读到 Record 对象中，如果读入失败，将游戏者的名字都设置为"匿名"，成绩纪录都设置为 9999。

当扫雷成功后，将当前用时与排行榜纪录中的用时进行比较，如果当前用时比较小，则出现一个对话框，提示用户输入名字，然后将名字和成绩保存到 Record 对象中。

1. 读入排行榜数据

在 MineFrame 类中定义 Record 属性，用来保存排行榜数据。

```
private Record record;
```

然后添加 getRecord() 和 setRecord() 方法，代码如下：

```
public Record getRecord() {
    return record;
}
public void setRecord(Record record) {
    this.record = record;
}
```

在 MineFrame 类中添加 initRecord() 方法，代码如下：

```
private void initRecord(){
    grade = Grade.LOWER;
    RecordDao rd = new RecordDao();
    record =rd.readRecord();
    if(record==null){ //读出失败
        record = new Record();
        record.setLowerName("匿名");
        record.setLowerScore(9999);
        record.setMedialName("匿名");
        record.setMedialScore(9999);
        record.setHigherName("匿名");
        record.setHigherScore(9999);
    }
}
```

首先将游戏难度级别 grade 设置为初级,然后读取排行榜数据,如果读取失败,将排行榜数据设置为指定的值。

在类 MineFrame 构造方法的第 4 行调用 initRecord()方法,代码如下:
```
public MineFrame(){
    createMenu();
    createUpPanel();
    initParameter(10,10,10);
    initRecord();
    …
}
```

2. 记录排行榜数据

扫雷成功后,如果本次游戏用时小于原来的纪录,则保存新成绩,在 MinePanel 类的 wins()方法中加入保存新纪录的代码。修改后的 wins()方法如下:
```
public void wins() {
    int second = utt.getSecond();      //获取游戏消耗的时间
    utt.cancel();
    mf.setGamming(false);
    mf.setStoped(true);
    JOptionPane.showMessageDialog(this, "恭喜,扫雷成功!");
    int grade = mf.getGrade();
    Record record = mf.getRecord();
    boolean newRecord = false;
    switch(grade){
    case Grade.LOWER:
        if(second < record.getLowerScore()){
            newRecord = true;
        }
        break;
    case Grade.MEDIAL:
        if(second < record.getMedialScore()){
            newRecord = true;
        }
        break;
    case Grade.HIGHER:
        if(second < record.getHigherScore()){
            newRecord = true;
        }
        break;
    }
    if(newRecord){
        String name = null;
        DialogRecordName dlg = new DialogRecordName(mf,second, grade);
        int option = dlg.openDialog();
        if(option==1){
            name = dlg.name.getText();
```

```java
            switch(grade){
            case Grade.LOWER:
                record.setLowerName(name);
                record.setLowerScore(second);
                break;
            case Grade.MEDIAL:
                record.setMedialName(name);
                record.setMedialScore(second);
                break;
            case Grade.HIGHER:
                record.setHigherName(name);
                record.setHigherScore(second);
                break;
            }
            RecordDao rd = new RecordDao();
            rd.writeRecord(record);
        }
    }
}
```

首先获取游戏所用的时间，然后根据当前的难度级别将游戏时间与相应难度级别的纪录进行比较，如果游戏用时较少，则产生新的纪录，此时打开输入名字对话框，将游戏者的名字和用时保存到相应难度级别的记录中，再将排行榜纪录保存到文件。

3. 显示排行榜数据

在菜单监听器处理"排行榜"菜单项的代码分支中添加以下代码：

```java
else if(mi.equals(menuItems[4])){
    DialogShowRecord dlg = new DialogShowRecord(MineFrame.this);
}
```

创建一个排行榜 DialogShowRecord 类的对象，由于 DialogShowRecord 类的构造方法中有语句 setVisible(true)，因此对话框创建后会自动显示。

1.8 附加功能

到上一节为止，扫雷功能已基本实现，在此基础上我们还可以添加一些辅助功能，比如增加声音效果、增加踩中地雷后产生的爆炸效果等。本节添加声音效果，有兴趣的读者可以根据自己的兴趣添加其他效果。

1.8.1 添加 sound()方法

在 MinePanel 类中添加播放音频的 sound()方法，代码如下：

```java
public void sound(String fileName){
    try {
        InputStream in = new FileInputStream(fileName);
        AudioStream as = new AudioStream(in);
        AudioPlayer.player.start(as);        //开始播放
```

```
        }catch(FileNotFoundException e){
        }
        catch(IOException e){
        }
    }
```

参数 fileName 指定要播放音频文件的文件名。

AudioStream 类和 AudioPlayer 类是 sun.audio 包中的类,要将这两个类导入进来,代码如下:

```
import sun.audio.AudioPlayer;
import sun.audio.AudioStream;
```

由于 sun 包并不是 API 公开接口的一部分,调用 sun 包的程序并不能确保工作在所有 Java 平台上,因此在使用 sun 包里的类时,可能会不成功。如果程序不能通过编译,先将项目的 jre 库删除,再重新添加进来,通常可以解决这一问题。方法是先选中扫雷项目,再选择 Project→Properties 菜单项,出现 Properties 对话框,在对话框左侧选中 Java Build Path,在右侧将 jre 库删除,然后再添加进来。

1.8.2 准备音频文件

找 3 个适当的声音文件(.wav 文件),将名字分别改为 success.wav、explode.wav 和 openblocks.wav,分别是在扫雷成功、扫雷失败和一次翻开多个小方块时播放的音频文件。在扫雷 project 文件夹中创建子文件夹 wav,将这 3 个文件复制到该文件夹中。

1.8.3 播放音频文件

在扫雷成功、失败、翻开多个方块时播放相应的音频文件。在扫雷成功时播放 success.wav 文件,在 wins()方法的第一行加入以下语句:

```
sound("wav/success.wav");
```

在扫雷失败时播放 explode.wav 文件,在 lose()方法的第一行加入以下语句:

```
sound("wav/explode.wav");
```

在 MinePanel 类中找到 open()方法,在小方块周围雷数为 0 的分支加入以下语句:

```
sound("wav/openblocks.wav");
```

声音处理完毕,重新运行程序,在扫雷成功、扫雷失败、一次翻开多个小方块时可以听到不同的声音。

1.9 作业

1. RecordDao 类中,在读写文件时,如果文件不存在或文件夹不存在,都会产生异常。我们的程序中并没有很好地处理这些异常。修改程序,处理这些异常。

2. 在自定义雷区对话框中,我们的程序没有对雷数、行列数进行限定,实际上这些值都应该有一个合理的范围,修改程序将这些值限定在某个范围内。

3. 在读取排行榜时,如果文件夹不存在,会产生异常,请修改程序,处理这个异常。

实训 2　网络五子棋

网络五子棋可以实现两个人在不同的客户端下棋。客户端界面如图 2.1 所示，棋盘占了界面的大部分空间，最下方是一些控制按钮。右侧分为上中下三部分，最上面用于显示用户头像、用户名和计时，中间用于显示已经连接到服务器的所有客户端用户列表，图 2.1 中有两个用户（Player1 和 Player2）连接到服务器上，下方用于显示一些实时信息。

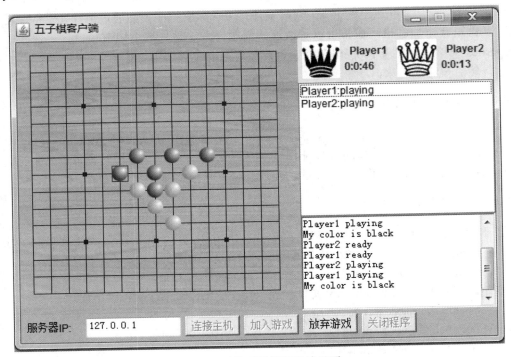

图 2.1　网络五子棋客户端界面

单击"连接主机"按钮连接服务器，客户端连接服务器后，可以看到所有已经连接到服务器的客户端（界面右侧中间部分显示客户端用户列表）。这时可以从列表中选择一个用户，单击"加入游戏"按钮，向该用户发起下棋申请，对方同意后随机猜先，开始下棋。猜到黑棋方先下子，并开始计时。在下棋过程中，如果某一方时间用完则判为输棋，下棋过程中也可以单击"放弃游戏"按钮认输。

服务器端界面比较简单，如图 2.2 所示。窗口顶端用一个标签显示当前连接到服务器上的客户端个数。中间用于显示有关信息，如果有客户端连接时，就显示该客户端的信息；如果客户端的状态改变后，则显示客户端的最新状态；用户下棋后，显示客户端的下棋位置等。最下面是一个按钮，用于关闭服务器。

设计棋子类和棋盘类，为了能够将棋盘和棋子显示出来，在第 1 节中先完成一个单机版的五子棋游戏，从第 2 节开始具体介绍网络五子棋的实现过程。

实训 2 网络五子棋

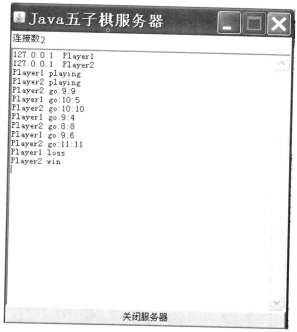

图 2.2 网络五子棋服务器端界面

2.1 单机版五子棋游戏

单机版五子棋程序相当于提供一个棋盘，两个用户用鼠标轮流下棋。系统负责判断下棋的位置是否合法以及输赢。运行界面如图 2.3 所示。

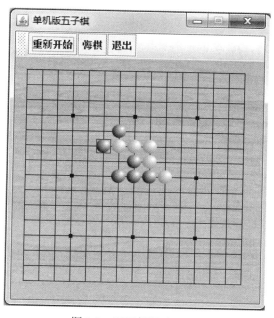

图 2.3 五子棋游戏界面

单机版五子棋的主要制作步骤包括创建窗口和工具栏、创建棋盘、创建棋子、实现单击鼠标下子、判断输赢、实现悔棋和重新开始等功能，以及改变鼠标的形状。

在实际的五子棋比赛中，为了公平起见，先行的黑棋是有禁手的，即某些棋型黑棋是不可以下的。为简单起见，本书中的五子棋程序暂不考虑禁手。

2.1.1 五子棋游戏窗口制作

五子棋窗口类从 JFrame 类继承，窗口上面包含一个工具栏，工具栏上有 3 个按钮，如图 2.4 所示。

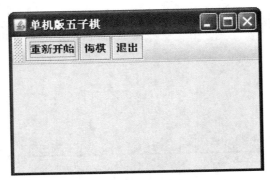

图 2.4 五子棋窗口

启动 Eclipse 后，创建一个名为 Five 的 Java Project，然后创建 Five 类（从 JFrame 类继承），在类中添加一个工具栏和 3 个按钮，在构造方法中创建这些组件对象，并将这些组件组织到窗口中。在主方法中创建一个 Five 类对象，代码如下：

```java
public class Five extends JFrame {
    private JToolBar toolbar;
    private JButton startButton,backButton,exitButton;
    public Five(){
        super("单机版五子棋");
        toolbar = new JToolBar();
        startButton = new JButton("重新开始");
        backButton = new JButton("悔棋");
        exitButton = new JButton("退出");
        toolbar.add(startButton);
        toolbar.add(backButton);
        toolbar.add(exitButton);
        this.add(toolbar, BorderLayout.NORTH);
        this.setBounds(200,200,300,200);
        this.setDefaultCloseOperation(EXIT_ON_CLOSE);
        this.setVisible(true);
    }
    public static void main(String[] args) {
        new Five();
    }
}
```

2.1.2 创建棋盘类

1. 准备图片

棋盘由一张背景图片和若干横线、竖线组成，棋盘效果如图 2.3 所示。

准备一张图片作为棋盘的背景，将图片的名字改为 board.jpg，在 project 文件夹中建立一个子文件夹 img，将背景图片文件放在其中。

2. 创建棋盘类

创建棋盘类 ChessBoard，该类从 JPanel 类继承，代码如下：

```
public class ChessBoard extends JPanel {
    public static final int MARGIN=15;        //边距
    public static final int SPAN=20;          //网格宽度
    public static final int ROWS=14;          //棋盘行数
    public static final int COLS=14;          //棋盘列数
    Image img;
    public ChessBoard(){
        img=Toolkit.getDefaultToolkit().getImage("img/board.jpg");
    }
    //画棋盘
    public void paintComponent(Graphics g){
        super.paintComponent(g);
        g.drawImage(img, 0, 0, this);
        for(int i=0;i<=ROWS;i++){              //画横线
            g.drawLine(MARGIN,    MARGIN + i*SPAN,
                       MARGIN + COLS*SPAN,    MARGIN + i*SPAN);
        }
        for(int i=0;i<=COLS;i++){              //画竖线
            g.drawLine(MARGIN + i*SPAN,    MARGIN,
                       MARGIN + i*SPAN,    MARGIN + ROWS*SPAN);
        }
        g.fillRect(MARGIN + 3*SPAN - 2, MARGIN + 3*SPAN - 2, 5, 5);
        g.fillRect(MARGIN + (COLS/2)*SPAN - 2, MARGIN + 3*SPAN - 2, 5, 5);
        g.fillRect(MARGIN + (COLS-3)*SPAN - 2, MARGIN + 3*SPAN - 2, 5, 5);
        g.fillRect(MARGIN + 3*SPAN - 2, MARGIN + (ROWS/2)*SPAN - 2, 5, 5);
        g.fillRect(MARGIN+ (COLS/2)*SPAN- 2, MARGIN+ (ROWS/2)*SPAN -2, 5, 5);
        g.fillRect(MARGIN+ (COLS-3)*SPAN- 2, MARGIN+ (ROWS/2)*SPAN- 2, 5, 5);
        g.fillRect(MARGIN + 3*SPAN - 2, MARGIN + (ROWS-3)* SPAN - 2, 5, 5);
        g.fillRect(MARGIN+ (COLS/2)*SPAN- 2, MARGIN+ (ROWS-3)*SPAN- 2, 5, 5);
        g.fillRect(MARGIN+ (COLS-3)*SPAN- 2, MARGIN+ (ROWS-3)*SPAN- 2, 5, 5);
    }
    public Dimension getPreferredSize(){
        return new Dimension(MARGIN*2+SPAN*COLS, MARGIN*2 +SPAN*ROWS);
    }
}
```

类中定义的常量分别是棋盘边框的宽度、网格的宽度，以及网格的行数和列数。注意，网格的行数比横线少 1，网格的列数比竖线少 1，将网格的行列数都定义为 14，则棋盘的横线和竖线都是 15 条。

当一个组件需要重画时，会调用到自身的 paintComponent()方法，因此我们在这个方法中绘制棋盘。首先将背景图片显示出来，然后第一个循环画出横线，第二个循环画出竖线，最后九行代码画出棋盘上的九个小的黑色正方形。

方法 getPreferredSize()返回棋盘最适合的尺寸，在后面的程序中确定主框架窗口大小时，会用到这个方法。

3. 显示棋盘

在 Five 类中添加 ChessBoard 类的属性成员。

  private ChessBoard boardPanel;

在构造方法中创建 boardPanel 对象，并将其添加到窗口框架中。

  boardPanel=new ChessBoard();
  this.add(boardPanel, BorderLayout.CENTER);

将原来设置窗口位置和大小的语句替换成以下 3 行代码：

  this.setLocation(200,200);
  this.pack();
  this.setResizable(false);

setLocation()方法将窗口的左上角设置到(200,200)。调用方法 pack()，使窗口的大小根据窗口中组件的大小自动调整，因此需要知道棋盘的大小，前面已经在棋盘类中添加了一个方法 getPreferredSize()，正是为了这一目的。

调用方法 setResizable(false)，使窗口的大小不可改变。

2.1.3 创建棋子类

1. 棋子类

棋子类的属性包括棋子的直径、颜色（黑或白）、在棋盘中的行列号，以及棋子所在的棋盘。棋子的直径是一个不变的值，因此定义为常量，将其值设置为比单元格宽度稍小一点。

棋子类的方法包括构造方法、属性的 get 方法，最主要的一个是将棋子画出来的 draw()方法。

程序代码如下：

```java
public class Chess {
    public static final int DIAMETER=ChessBoard.SPAN-2;
    private int col;        //棋子在棋盘中的 x 索引
    private int row;        //棋子在棋盘中的 y 索引
    private Color color;    //颜色
    ChessBoard cb;
    public Chess(ChessBoard cb,int col,int row,Color color){
        this.cb = cb;
        this.col=col;
        this.row=row;
        this.color=color;
    }
    public int getCol() {
        return col;
```

```
        }
        public int getRow() {
            return row;
        }
        public Color getColor() {
            return color;
        }
        public void draw(Graphics g){
            int xPos= col * cb.SPAN + cb.MARGIN;
            int yPos= row * cb.SPAN + cb.MARGIN;
            Graphics2D g2d = (Graphics2D) g;
            RadialGradientPaint paint=null;
            int x = xPos + DIAMETER/4;
            int y = yPos - DIAMETER/4;
            float[] f = {0f, 1f};
            Color[] c = {Color.WHITE, Color.BLACK};
            if(color==Color.black){
                paint = new RadialGradientPaint(x,y, DIAMETER, f, c );
            }
            else if(color==Color.white){
                paint = new RadialGradientPaint(x, y, DIAMETER*2, f, c);
            }
            g2d.setPaint(paint);
            //以下两行使边界更均匀
            g2d.setRenderingHint(RenderingHints.KEY_ANTIALIASING,
                            RenderingHints.VALUE_ANTIALIAS_ON);
            g2d.setRenderingHint(RenderingHints.KEY_ALPHA_INTERPOLATION,
                            RenderingHints.VALUE_ALPHA_INTERPOLATION_DEFAULT);
            Ellipse2D e = new Ellipse2D.Float(xPos-DIAMETER/2, yPos-DIAMETER/2,
                            DIAMETER,DIAMETER);
            g2d.fill(e);
        }
    }
```

为了使棋子有立体效果，我们使用圆形辐射颜色渐变模式填充方式。下面结合图 2.5 介绍黑棋棋子的绘制过程。

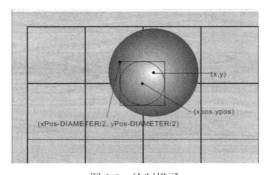

图 2.5　绘制棋子

图中的小圆表示要绘制的黑棋棋子，其直径是 DIAMETER，先获取棋子的中心坐标(xPos, yPos)。图中的大圆是用于填充的，其直径为 2*DIAMETER，中心为棋子的高光点，位于棋子右上方 1/4 处。

RadialGradientPaint 类提供了圆形辐射颜色渐变模式填充方式，在我们的程序中使用 5 个参数的构造方法构造一个 RadialGradientPaint 对象，前两个参数为填充的中心坐标，第 3 个参数是渐变圆的半径，第 4 个参数是实型数组，在我们的程序中是{0,1}，第 5 个参数是 Color 型数组，在我们的程序中是{Color.WHITE, Color.BLACK}。第 4 个参数和第 5 个参数要配合使用，元素的个数要相同，0 表示渐变的中心位置，对应颜色 Color.WHITE，1 表示圆周上，对应的颜色是 Color.BLACK，中间则是两种颜色的逐渐过渡，当然这两个数组也可以有多个元素，实现多种颜色的渐变。

创建 RadialGradientPaint 对象后，调用 setPaint()方法将其应用到 g2d 中，最后创建棋子椭圆并填充。注意 Ellipse2D 是抽象类，Ellipse2D.Float 是 Ellipse2D 的一个内部子类，其构造方法的参数分别是椭圆外切矩形的左上角坐标和外切矩形的长和宽。

调用 setRenderingHint()方法的两行代码的作用是使棋子的边界绘制得更平滑。

白棋的绘制与此类似，可以仔细分析程序中创建 RadialGradientPaint 对象的代码，理解白棋棋子的绘制。

2. 在棋盘上画出棋子

为了测试棋子类，我们在棋盘类的 paintComponent()方法的最后加上以下 4 行代码：

　　Chess c1 = **new** Chess(**this**,2,1,Color.*BLACK*);
　　Chess c2 = **new** Chess(**this**,5,2,Color.*WHITE*);
　　c1.draw(g);
　　c2.draw(g);

上述代码创建两个棋子 c1（2 列、1 行、黑棋）和 c2（5 列、2 行、白棋），并将两个棋子绘制出来，效果如图 2.6 所示。

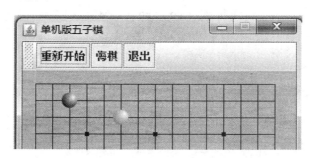

图 2.6　棋盘上的两个棋子

2.1.4　实现单击鼠标下棋

前面我们在棋盘类的 paintComponent()方法中创建了两个棋子并显示出来，而在实际的对局中是通过单击鼠标下棋的，下面的程序实现单击鼠标下棋。

为了记录下棋的过程，在棋盘类中增加以下 3 个属性：

　　Chess[] chessList =new Chess[(ROWS+1)*(COLS+1)] ;　　//记录棋盘上的棋子的数组
　　int chessCount = 0;　　//当前棋盘上棋子的个数

```
boolean isBlack=true;        //下一步轮到哪一方下棋，默认开始时是黑棋先
```
在某一位置下棋之前，应该先检测该位置是否已经有了棋子，如果已经有了棋子，则不可以在该位置再下棋，因此为棋盘类添加一个方法 hasChess()，代码如下：
```
private boolean hasChess(int col,int row){
    for(int i = 0; i< chessCount; i++){
        Chess ch = chessList[i];
        if(ch!=null&&ch.getCol()==col&&ch.getRow()==row)
            return true;
    }
    return false;
}
```
在棋子数组中逐个判断棋子是否在指定的位置，如果找到一个棋子在该位置，返回 true，否则返回 false。

为了响应鼠标消息，给棋盘类添加一个内部类 MouseMonitor，该类从 MouseAdapter 类继承，在类中重写函数 mousePressed()，当鼠标按下时会调用此方法，代码如下：
```
class MouseMonitor extends MouseAdapter{
    public void mousePressed(MouseEvent e){
        //将鼠标点击的像素坐标转换成网格索引
        int col=(e.getX()-MARGIN+SPAN/2)/SPAN;
        int row=(e.getY()-MARGIN+SPAN/2)/SPAN;
        //落在棋盘外不能下棋
        if(col<0||col>COLS||row<0||row>ROWS)
            return;
        //如果 x，y 位置已经有棋子存在，不能下棋
        if(hasChess(col, row))
            return;
        Chess ch=new Chess(ChessBoard.this, col, row, isBlack?Color.black:Color.white);
        chessList[chessCount++]=ch;
        repaint();        //通知系统重新绘制
        isBlack=!isBlack;
    }
}
```
首先将鼠标的像素坐标转换为棋盘坐标，在图 2.7 所示的"矩形 1"区域内按下鼠标，其棋盘坐标应该是(2,1)。先将按下鼠标点的行列坐标都加上单元格宽度的一半，这样的坐标范围相当于"矩形 2"的区域，再将其整除单元格的宽度，就是我们需要的棋盘坐标。

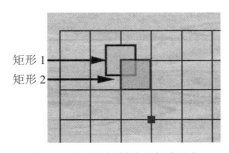

图 2.7　像素坐标转换为棋盘坐标

然后判断该位置是否可以下棋，如果鼠标位于棋盘之外或该位置已经有了棋子，则不能下棋，直接返回。如果可以下棋，则创建一个棋子对象，棋子的颜色由属性 isBlack 当前的值确定，如果 isBlack 为真，则创建黑子，否则创建白子。将棋子添加到棋子数组中，再将棋盘上的棋子数增加 1。

最后调用 repaint()方法，重绘棋盘，再将 isBlack 的值翻转一次，实现黑白两方轮流下棋。

将 paintComponent()方法的最后 4 行用下面的程序替换。通过循环将棋盘上的所有棋子绘制出来。为了标识最后一步棋的位置，我们在最后的棋子上画一个红色矩形。

```
for(int i=0;i<chessCount;i++){
    chessList[i].draw(g);
    if(i==chessCount-1){//如果是最后一个棋子
        int xPos=chessList[i].getCol()*SPAN+MARGIN;
        int yPos=chessList[i].getRow()*SPAN+MARGIN;
        g.setColor(Color.red);
        g.drawRect(xPos-Chess.DIAMETER/2, yPos-Chess.DIAMETER/2,
            Chess.DIAMETER, Chess.DIAMETER);
    }
}
```

在棋盘类的构造方法中注册鼠标监听事件，完成通过鼠标下棋的功能。

```
this.addMouseListener(new MouseMonitor());
```

以上程序已经实现了黑白两方轮流下棋的功能，下面一节中将实现判断输赢的功能。

2.1.5 判断赢棋

每下一个棋子都要判断是否有五个自己的棋子连成一线了，一旦五个棋子连成一线就赢棋，棋局结束。

在判断五子连成一线时，要考虑四个方向：水平、垂直、从左上角到右下角、从左下角到右上角，如图 2.8 所示。

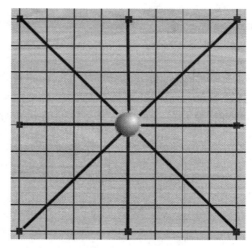

图 2.8　四个方向判断五子连成一线

例如，在图 2.8 中间下了一个黑子，判断水平方向黑子是否连成一线，方法是先检查中间左侧的点是否有黑子,如果有黑子,则继续向左判断,直到不是黑子或者已经够五个黑子了,则结束;否则再判断中间右侧的点是否有黑子,如果有黑子,继续向右判断,直到不是黑子或已经够五个黑子,则结束。

因此我们要在棋盘类中再添加一个方法 hasChess()(与前一节的 hasChess()方法是重载关系),用来判断某个点有没有黑子或有没有白子。

```java
private boolean hasChess(int col, int row, Color color){
    for(int i=0; i<chessCount; i++){
        Chess ch = chessList[i];
        if(ch!=null&&ch.getCol()==col&&ch.getRow()==row &&ch.getColor()==color)
            return true;
    }
    return false;
}
```

函数通过一个循环,逐个判断棋盘上的棋子是不是参数指定位置和颜色的棋子,如果是,返回 true,如果循环结束仍然没有找到参数指定的棋子,则返回 false。

判断胜负的方法如下:

```java
private boolean isWin(int col, int row){
    int continueCount=1;      //连续棋子的个数
    Color c=isBlack?Color.black:Color.white;
    //横向向左寻找
    for(int x=col-1;x>=0;x--){
        if(hasChess(x,row,c))
            continueCount++;
        else
            break;
    }
    //横向向右寻找
    for(int x=col+1;x<=COLS;x++){
        if(hasChess(x,row,c))
            continueCount++;
        else
            break;
    }
    if(continueCount>=5)
        return true;
    else
        continueCount=1;
    //继续另一种情况的搜索:纵向
    //向上搜索
    for(int y=row-1;y>=0;y--){
        if(hasChess(col,y,c))
            continueCount++;
        else
            break;
    }
```

```java
//纵向向下寻找
for(int y=row+1;y<=ROWS;y++){
    if(hasChess(col,y,c))
        continueCount++;
    else
        break;
}
if(continueCount>=5)
    return true;
else
    continueCount=1;
//继续另一种情况的搜索：右上到左下
//向右上寻找
for(int x=col+1,y=row-1; y>=0&&x<=COLS; x++,y--){
    if(hasChess(x,y,c))
        continueCount++;
    else
        break;
}
//向左下寻找
for(int x=col-1,y=row+1; x>=0&&y<=ROWS; x--,y++){
    if(hasChess(x,y,c))
        continueCount++;
    else
        break;
}
if(continueCount>=5)
    return true;
else
    continueCount=1;
//继续另一种情况的搜索：左上到右下
//向左上寻找
for(int x=col-1,y=row-1; x>=0&&y>=0; x--,y--){
    if(hasChess(x,y,c))
        continueCount++;
    else
        break;
}
//向右下寻找
for(int x=col+1,y=row+1; x<=COLS&&y<=ROWS; x++,y++){
    if(hasChess(x,y,c))
        continueCount++;
    else
        break;
}
if(continueCount>=5)
```

```
                return true;
        else
                return false;
}
```

我们以横向是否有五个相同的棋子连成一线为例，介绍判断的过程，其他方向的判断类似。

首先得到当前下棋棋子的颜色 c，再给记录连续棋子数的变量赋值为 1。从当前下棋位置的左面一列开始，如果该位置有 c 颜色的棋子，则将连续的棋子数加 1，继续向左，否则结束循环。再从当前下棋位置的右面一列开始，继续向右判断。循环结束后，如果连续的棋子数大于等于 5，则赢棋，返回 true。否则将连续的棋子数重新设置为 1，再判断其他方向是否连成五子一线。

如果四个方向都判断完，仍然没有五子连成一线，则返回 false。

修改监听器类的 mousePressed() 方法，每下一步棋都要判断是否赢棋，如果赢棋则棋局结束，进行结束处理。由于棋局结束后，按下鼠标不能再产生棋子了，因此我们在棋盘类中添加一个成员变量 isGamming，表示是否为下棋状态，并将其初始化为 true。

```
boolean isGamming=true;//是否正在游戏
```

在 mousePressed() 方法的最开始添加如下语句，如果不是正在游戏，则返回：

```
if(!isGamming) return;
```

在 mousePressed() 方法中 "repaint();" 语句之后添加如下语句：

```
if(isWin(col, row)){
        String colorName=isBlack?"黑棋":"白棋";
        String msg=String.format("恭喜，%s 赢了！", colorName);
        JOptionPane.showMessageDialog(ChessBoard.this, msg);
        isGamming=false;
}
```

如果赢棋，则出现一个信息框，显示祝贺赢棋的信息，然后将 isGamming 设置为 false。

2.1.6 实现工具栏上的功能

实现工具栏上 "重新开始" "悔棋" 和 "退出" 三个按钮的功能。

首先在棋盘类中添加两个方法 restartGame() 和 goback()，分别实现重新开始和悔棋的功能。

重新开始需要做的事情就是清空棋子、将棋盘上的棋子数设置为 0、重绘棋盘、将 isBlack 和 isGamming 设置为 true，代码如下：

```
public void restartGame(){
        //清除棋子
        for(int i=0;i<chessList.length;i++){
                chessList[i]=null;
        }
        //恢复游戏相关的变量值
        isBlack=true;
        isGamming=true;        //游戏是否结束
        chessCount =0;         //当前棋盘棋子个数
        repaint();
}
```

悔棋功能需要做的事情就是将最后下的一个棋子去掉，将棋盘上的棋子数减 1，将 isBlack 的值反转，重绘棋盘。

```java
public void goback(){
    if(chessCount==0)          //如果棋子数是 0，则不能悔棋
        return ;
    chessList[chessCount-1]=null;
    chessCount--;
    isBlack=!isBlack;
    repaint();
}
```

然后在 Five 类中添加内部监听器类 ActionMonitor。

```java
class ActionMonitor implements ActionListener{
    public void actionPerformed(ActionEvent e) {
        if(e.getSource()==startButton){
            boardPanel.restartGame();
        }
        else if(e.getSource() == backButton){
            boardPanel.goback();
        }
        else if(e.getSource() == exitButton){
            System.exit(0);
        }
    }
}
```

如果单击的是"重新开始"按钮，则调用棋盘类中的 restartGame()方法，开始新的棋局；如果单击的是"悔棋"按钮，则调用棋盘类中的 goback()方法，实现悔棋功能；如果单击的是"退出"按钮，则退出程序。

在 Five 类的构造方法中为三个按钮注册监听器，代码如下：

```java
ActionMonitor monitor = new ActionMonitor();
startButton.addActionListener(monitor);
backButton.addActionListener(monitor);
exitButton.addActionListener(monitor);
```

经过以上处理，目前的程序已经可以实现自动判断输赢的功能了，也就是说实现了两个人使用鼠标轮流下棋的功能，并可以判断输赢。

2.1.7 改变鼠标的形状

下面的程序将实现通过鼠标的形状来表示某点是否可以下棋，当鼠标在棋盘内部移动时，如果该位置没有棋子，可以下子，将鼠标形状设置为手形光标；如果该位置不能下子，将鼠标形状设置为标准箭头光标。因此要监听鼠标的移动消息，需要在棋盘类中添加内部类 MouseMotionMonitor，代码如下：

```java
class MouseMotionMonitor extends MouseMotionAdapter{
    public void mouseMoved(MouseEvent e){
        int col=(e.getX()-MARGIN+SPAN/2)/SPAN;
```

```
            int row =(e.getY()-MARGIN+SPAN/2)/SPAN;
            if(col<0||col>ROWS||row<0||row>COLS||!isGamming||hasChess(col,row))
                    ChessBoard.this.setCursor(new Cursor(Cursor.DEFAULT_CURSOR));
            else
                    ChessBoard.this.setCursor(new Cursor(Cursor.HAND_CURSOR));
        }
    }
```

在 mouseMoved()方法中，首先计算鼠标位置的棋盘坐标，如果处于棋盘之外、或不是处于下棋状态、或该位置已经有棋子，则将鼠标光标设置为标准箭头形，否则将鼠标光标设置为手形。

然后在棋盘类的构造方法中注册监听器。

```
        this.addMouseMotionListener(new MouseMotionMonitor());
```

至此，单机版五子棋程序已经完成。

2.2 服务器端界面制作

上面一节我们完成了单机版五子棋的设计，所设计的棋盘类和棋子类可以直接应用到网络版五子棋中。在后面的几节中将逐步实现网络版五子棋的功能，本节先设计网络版五子棋的服务器端界面。

首先创建一个名为 NetFive 的 Project，然后创建 FiveServer 类，代码如下：

```
public class FiveServer extends Frame implements ActionListener{
    Label lStatus=new Label("当前连接数： ",Label.LEFT);
    TextArea taMessage=new TextArea("",22,50,TextArea.SCROLLBARS_VERTICAL_ONLY);
    Button btServerClose=new Button("关闭服务器");
    public static void main(String[] args){
        FiveServer fs = new FiveServer();
    }
    public FiveServer(){
        super("Java 五子棋服务器");
        btServerClose.addActionListener(this);
        add(lStatus,BorderLayout.NORTH);
        add(taMessage,BorderLayout.CENTER);
        add(btServerClose,BorderLayout.SOUTH);
        setLocation(400,100);
        pack();
        setVisible(true);
        setResizable(false);
    }
    public void actionPerformed(ActionEvent e) {
        if(e.getSource()==btServerClose){
            System.exit(0);
        }
    }
}
```

FiveServer 类从 Frame 类继承，并实现了 ActionListener 接口，因此类中实现了 actionPerformed 方法，单击"关闭服务器"按钮可以关闭服务器程序。类中定义了服务器窗口中的三个组件，在构造方法中将三个组件组装到窗口中，并为按钮注册监听器，实现程序的退出功能。

FiveServer 类包含一个主方法。在主方法中，创建一个 FiveServer 对象。运行 FiveServer 类的结果与图 2.2 所示相同。

2.3 创建客户端界面

客户端界面整体使用 BoardLayout 布局，分成三个区域（见图 2.1），第一个区域是棋盘，组装在窗口的中间，第二个区域包括计时区、用户列表区和信息区，组装到窗口的右侧，第三个区域是控制按钮区，组装在窗口的下方。

第二个区域又包括三个区域，同样使用 BoardLayout 布局，按上中下的位置排列。

为了清晰，我们分别为这些区域创建棋盘类 PanelBoard、计时面板类 PanelTiming、用户列表类 PanelUserList、信息面板类 PanelMessage 和控制按钮面板类 PanelControl。

2.3.1 创建主窗口和棋盘

将单机版的棋盘类和棋子类复制过来，将棋盘类的名字改为 PanelBoard，并将属性 isGamming 初始化为 false。

   boolean isGamming=false; //是否正在游戏

删除棋盘类中的 restartGame()方法和 goback()方法。

与单机版一样建立 img 文件夹，将棋盘背景图片保存在该文件夹中，同时再找两个图片分别表示黑棋用户头像和白棋用户头像，如图 2.9 所示。

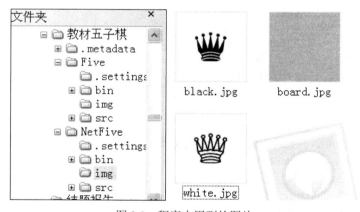

图 2.9 程序中用到的图片

创建客户端窗口类 FiveClient，代码如下：

```
public class FiveClient extends Frame {
    PanelBoard board;
    public static void main(String[] args) {
        FiveClient fc = new FiveClient();
    }
```

```
        public FiveClient(){
            super("五子棋客户端");
            board=new PanelBoard();
            this.add(board, BorderLayout.CENTER);
            this.setLocation(300,100);
            pack();
            this.setResizable(false);
            this.setVisible(true);
        }
    }
```

在 FiveClient 类中定义一个棋盘类的对象属性，在构造方法中创建棋盘类对象，并组装到窗口的中间。

运行 FiveClient 类的结果可参看前面图 2.1 的棋盘区域。

2.3.2 创建客户端界面右侧的三个类

棋盘右侧包括计时面板类 PanelTiming、用户列表类 PanelUserList、信息面板类 PanelMessage。

1. 创建 PanelTiming 类

计时面板由 6 个组件组成：自己的头像、名字、剩余时间和对手的头像、名字、剩余时间，分别用 6 个 JLabel 组件实现。代码如下：

```
        public class PanelTiming extends JPanel {
            JLabel myIcon;
            JLabel opIcon;
            JLabel myName;
            JLabel opName;
            JLabel myTimer;
            JLabel opTimer;
            Icon blackIcon;
            Icon whiteIcon;
            public PanelTiming() {
                blackIcon = new ImageIcon("img\\black.jpg");
                whiteIcon = new ImageIcon("img\\white.jpg");
                this.myIcon = new JLabel(blackIcon,SwingConstants.CENTER);
                this.opIcon = new JLabel(whiteIcon,SwingConstants.CENTER);
                this.myName = new JLabel("My   Name",SwingConstants.CENTER);
                this.opName = new JLabel("Op   name",SwingConstants.CENTER);
                this.myTimer = new JLabel("00:00:00");
                this.opTimer = new JLabel("00:00:00");
                JPanel myText = new JPanel(new GridLayout(2,1));
                myText.add(myName);
                myText.add(myTimer);
                JPanel opText = new JPanel(new GridLayout(2,1));
                opText.add(opName);
                opText.add(opTimer);
                this.add(myIcon);
```

```java
            this.add(myText);
            this.add(opIcon);
            this.add(opText);
        }
        public void setMyIcon(String color) {
            if(color.equals("black")){
                this.myIcon.setIcon(blackIcon);
            }
            else{
                this.myIcon.setIcon(whiteIcon);
            }
        }
        public void setOpIcon(String color) {
            if(color.equals("black")){
                this.opIcon.setIcon(blackIcon);
            }
            else{
                this.opIcon.setIcon(whiteIcon);
            }
        }
        public void setMyName(String name) {
            this.myName.setText(name);
        }
        public void setOpName(String name) {
            this.opName.setText(name);
        }
        public void setMyTime(int time) {
            int h = time/3600;
            int m = (time - h*3600)/60;
            int s = time%60;
            this.myTimer.setText(h+":"+m+":"+s);
        }
        public void setOpTime(int time) {
            int h = time/3600;
            int m = (time - h*3600)/60;
            int s = time%60;
            this.opTimer.setText(h+":"+m+":"+s);
        }
    }
```

在构造方法中，创建所需要的组件，先将自己的用户名和计时组装到 myText 中，将对手的用户名和计时组装到 opText 中，然后将 myIcon、myText、opIcon 和 opText 以流式布局组装到计时面板中。

在我们的程序中，用户名是在客户端连接到服务器后由服务器给起名，因此需要方法 setMyName() 设置自己的用户名，而对手的用户名是在猜先后才能知道，因此也需要设置对手用户名的方法 setOpName()。自己是黑棋还是白棋也是在猜先后才知道的，需要两个分别用来设置自己头像和对手头像的 setMyIcon()方法和 setOpIcon()方法。下棋后开始计时，自己和对手的剩余时间要不停地改变，需要两个方法 setMyTime()和 setOpTime()来分别设置自己的剩余时间和对手的剩余时间。

设置剩余时间方法的参数是以秒为单位的，而显示的格式为"时:分:秒"，因此方法中要将秒转换为"时:分:秒"的格式。

2. 创建 PanelUserList 类和 PanelMessage 类

PanelUserList 类和 PanelMessage 类都比较简单。PanelUserList 类只包含一个 List 组件，PanelMessage 类只包含一个 TextArea 组件。

PanelUserList 类的代码如下：

```
public class PanelUserList extends Panel {
    public List userList = new List(8);
    public PanelUserList () {
        setLayout (new BorderLayout ());
        add (userList, BorderLayout.CENTER);
    }
}
```

PanelUserList 类中创建一个 8 行的 List 组件，并将 List 组件放在 PanelUserList 的中间。

PanelMessage 类的代码如下：

```
public class PanelMessage extends Panel {
    public TextArea mesageArea;
    public PanelMessage () {
        setLayout (new BorderLayout ());
        mesageArea=new TextArea("",12,20,TextArea.SCROLLBARS_VERTICAL_ONLY);
        add (mesageArea, BorderLayout.CENTER);
    }
}
```

PanelMessage 类中创建一个 12 行 20 列的文本域，并包含垂直方向的滚动条。

3. 将三个组件组装到主界面中

在 FiveClient 类中添加如下三个属性：

```
PanelUserList userList;
PanelMessage message;
PanelTiming timing;
```

在构造方法中创建这三个对象，并将它们组装到 east 中，然后再将 east 组装到主窗口的右侧，代码如下：

```
timing = new PanelTiming();
userList = new PanelUserList();
message = new PanelMessage();
Panel east = new Panel();
east.setLayout(new BorderLayout());
east.add(userList, BorderLayout.CENTER);
east.add(message, BorderLayout.SOUTH);
east.add(timing, BorderLayout.NORTH);
this.add(east, BorderLayout.EAST);
```

程序运行后的结果与图 2.1 相比，除了没有最下面的控制按钮面板，其他部分都齐了。

2.3.3 创建客户端界面下方的控制面板类

1. 创建控制面板类

控制面板也比较简单，都是由基本组件组成的，包括标签、文本框和按钮，以流式布局的方式从左到右排列。代码如下：

```
public class PanelControl extends Panel {
    public Label IPlabel = new Label("服务器 IP：", Label.LEFT);
    public TextField inputIP = new TextField ("127.0.0.1", 12);
    public Button connectButton = new Button ("连接主机");
    public Button joinGameButton = new Button ("加入游戏");
    public Button cancelGameButton = new Button("放弃游戏");
    public Button exitGameButton = new Button ("关闭程序");
    //构造函数，负责 Panel 的初始布局
    public PanelControl () {
        setLayout (new FlowLayout (FlowLayout.LEFT));
        setBackground (new Color (200,200,200));
        add (IPlabel);
        add (inputIP);
        add (connectButton);
        add (joinGameButton);
        add (cancelGameButton);
        add (exitGameButton);
    }
}
```

构造方法中调用 setBackground()方法来设置背景颜色。如果不设置，默认的颜色较浅，会与窗口内部颜色太接近。将 PanelControl 设置为流式布局，各组件依次组装到控制面板中。

2. 将 PanelControl 组装到客户端主窗口中

在 FiveClient 类中添加 PanelControl 类的对象属性。

```
PanelControl control;
```

在构造方法中创建 control 对象，并将它们组装到主窗口的下侧，代码如下：

```
control = new PanelControl();
this.add(control, BorderLayout.SOUTH);
```

至此，客户端界面已经完成，运行后与图 2.1 相同，只是还不能下棋。

2.4 实现"连接主机"按钮的功能

服务器运行后，如果有客户端连接，应该记录客户端的相关信息，包括用户名、客户端的 Socket 以及用户的状态等，因此应该设计一个类来保存这些信息。

客户端单击"连接主机"按钮，与服务器连接，连接成功后，服务器给客户端起用户名，并将用户名通知客户端，以便设置计时面板中自己的用户名。同时服务器还要将新登录的客户端信息发送给其他所有客户端，以便更新每个客户端的用户列表，也要将所有已经连接的客户

端信息发送给新连接的客户端，以便更新新连接客户端的用户列表。假设客户端 B 和客户端 C 已经连接到服务器上，当客户端 A 连接到服务器时，处理的流程如图 2.10 所示。

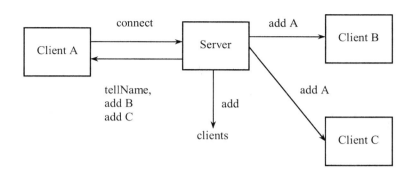

图 2.10　连接主机的流程

客户端 A 连接到服务器，服务器给客户端 A 分配名字并记录客户端 A 的信息（clients 是一个记录客户端信息的 ArrayList 链表），然后告知客户端 A 的用户名（发送 tellName 命令），再向客户端 B 和客户端 C 发送 add 命令，通知客户端 B 和客户端 C 加入客户端 A，同时向客户端 A 发送两个 add 命令，分别加入客户端 B 和客户端 C。

2.4.1　连接服务器获取用户名

1. 记录客户端信息的 Client 类

在 FiveServer 类中添加一个内部类 Client，用于记录有关连接到服务器上的客户端的信息。

```
class Client{
    String name;
    Socket s;
    String    state;        //1.ready    2.playing
    Client opponent;        //对手
    public Client(String name, Socket s) {
        this.name = name;
        this.s = s;
        this.state = "ready";
        this.opponent = null;
    }
}
```

四个属性分别是用户名、客户端的 Socket、状态（下棋中还是等待）和下棋的对手。当客户端连接成功后，其状态是准备好的状态，还没有下棋的对手，只有处于下棋状态时才有对手。

2. 为 FiveServer 类增加属性

在 FiveServer 类中添加一个 ServerSocket 属性 ss、一个表示五子棋程序使用的端口号的常量 TCP_PORT、一个用于记录当前连接到服务器上的客户端的数量 clientNum、一个用于为客户端起名的序号 clientNameNum 和一个保存客户端信息的链表 clients。

添加的属性如下:
```
ServerSocket ss = null;
public static final int TCP_PORT = 4801;
static int clientNum = 0;
static int clientNameNum = 0;
ArrayList<Client> clients = new ArrayList<Client>();
```

3. 为 FiveServer 类添加两个方法

在 FiveServer 类中添加 startServer()方法和 tellName()方法。

startServer()方法用于启动服务器,代码如下:

```
public void startServer(){
    try {
        ss = new ServerSocket(TCP_PORT);
        while(true){
            Socket s = ss.accept();
            clientNum++;
            clientNameNum++;
            Client c = new Client("Player"+clientNameNum, s);
            clients.add(c);
            lStatus.setText("连接数" + clientNum);
            String msg = s.getInetAddress().getHostAddress() +
                    "Player" + clientNameNum + "\n";
            taMessage.append(msg);
            tellName(c);
        }
    } catch (IOException e) {
        e.printStackTrace();
    }
}
```

在 startServer()方法中,首先创建 ServerSocket 对象,然后在循环中等待客户端的连接。一旦有客户端连接,将 clientNum 和 clientNameNum 的值增加 1,创建 Client 对象,并添加到客户端链表 clients 中。更新显示客户端连接数的标签组件,并将客户端的信息显示在信息区域。最后调用 tellName()方法通知客户端用户名。

tellName()方法的代码如下:

```
private void tellName(Client c) {
    DataOutputStream dos=null;
    try{
        dos = new DataOutputStream(c.s.getOutputStream());
        dos.writeUTF("tellName" +":" + c.name);
    } catch (IOException e) {
        e.printStackTrace();
    }
}
```

tellName()方法向对应的客户端发送消息,消息的第一个单词是 tellName,第二个单词是客户端的用户名，中间用冒号分隔。

在主方法中调用 startServer()方法，启动服务器。

```
public static void main(String[] args){
    FiveServer fs = new FiveServer();
    fs.startServer();
}
```

这样在运行服务器程序后，会自动启动服务器监听客户端的连接请求。

4. 为 FiveClient 类的按钮注册监听器

首先在 FiveClient 类中添加三个属性，分别表示自己的用户名、对手的用户名和当前是否连接到服务器的变量。

```
String myname;
String opname;
public boolean isConnected = false;
```

然后添加内部监听器类，代码如下：

```
class ActionMonitor implements ActionListener{
    public void actionPerformed(ActionEvent e) {
        if(e.getSource()==control.exitGameButton){
            System.exit(0);
        }
        else if(e.getSource() == control.connectButton){
            Socket s;
            String ip = control.inputIP.getText();
            try {
                s = new Socket(ip, FiveServer.TCP_PORT);
                isConnected  = true;
                InputStream is = s.getInputStream();
                DataInputStream dis = new DataInputStream(is);
                String msg = dis.readUTF();
                String[] words = msg.split(":");
                myname = words[1];
                userList.userList.add(myname + ":ready");
                timing.setMyName(myname);
                message.mesageArea.append("My name: " + myname + "\n");
                control.exitGameButton.setEnabled(true);
                control.connectButton.setEnabled(false);
                control.joinGameButton.setEnabled(true);
                control.cancelGameButton.setEnabled(false);
            } catch (UnknownHostException e1) {
                e1.printStackTrace();
            } catch (IOException e1) {
                e1.printStackTrace();
            }
        }
        else if(e.getSource() == control.cancelGameButton){
```

```
                else if(e.getSource() == control.joinGameButton){

                }
            }
        }
```

在 actionPerformed()方法中，首先判断是哪个按钮产生的事件，如果是"关闭程序"按钮，则退出程序；如果是"连接主机"按钮，则进行连接处理，另外两个按钮暂时未处理。

在连接处理代码中，通过创建 Socket 对象与服务器连接，连接后将 isConnected 设置为 true，然后读取从服务器发送来的 tellName 消息，将自己的名字等信息分别显示到计时面板、用户列表面板和消息面板中，最后设置四个按钮的状态。连接服务器后，"连接主机"按钮设置为不可用，"加入游戏"按钮设置为可用，"放弃游戏"按钮设置为不可用。

在客户端与服务器端通信时，命令中的各个单词间采用冒号分隔，例如 tellName:Player1。当然在后面的程序中也有三个单词的命令，可以使用 split()方法将这些单词从整个命令分离出来，放到一个字符串数组中。

在构造方法中为控制面板的按钮注册监听器，并设置按钮的初始状态。

```
ActionMonitor monitor = new ActionMonitor();
control.exitGameButton.addActionListener(monitor);
control.connectButton.addActionListener(monitor);
control.joinGameButton.addActionListener(monitor);
control.cancelGameButton.addActionListener(monitor);
control.exitGameButton.setEnabled(true);
control.connectButton.setEnabled(true);
control.joinGameButton.setEnabled(false);
control.cancelGameButton.setEnabled(false);
```

未连接服务器时，只有"连接主机"按钮和"关闭游戏"按钮可用。

2.4.2 将已经连接的客户端添加到用户列表中

新的客户端登录后应该将原来已经登录的客户端显示在自己的用户列表中，同样自己也应该出现在其他客户端的用户列表中。

1. 创建命令类 Command

在接下来的程序中，客户端与服务器之间要进行各种信息的传递，为了规范信息，把整个信息分成两段，第一部分是信息类别命令，如前面用到的 tellName，第二部分是参数，如 tellName 命令后面要有用户名，命令与参数之间用冒号分隔。有些命令没有参数，有些命令有一个参数，有些命令有两个参数，参数间也用冒号分隔。为了便于管理，我们创建一个 Command 类，在 Command 类中定义命令常量。

```
Public class Command {
    Public static final String TELLNAME = "tellname";
    Public static final String ADD = "add";
}
```

目前只有两个命令，其中 tellName 在前面已经使用过，add 将在下面的程序中使用，后面还会有很多命令，用到时再随时添加。

将 FiveServer 类中的 tellName()方法中的下面语句

  dos.writeUTF("tellName" +":" + c.name);

改为

  dos.writeUTF(Command.*TELLNAME* +":" + c.name);

2. 服务器向客户端告知已经连接的客户端

在 FiveServer 类中添加两个方法：addAllUserToMe()和 addMeToAllUser()，分别将所有已连接的客户端通知新连接的客户端和将新客户端通知给所有已经连接的其他客户端。

addAllUserToMe()方法通过循环，使用 add 命令将 clients 中的每个客户端发送到新的客户端，命令的格式是"add:用户名:状态"。代码如下：

```
private void addAllUserToMe(Client c) {
    DataOutputStream dos=null;
    for(int i=0; i<clients.size(); i++){
        if(clients.get(i) != c){
            try {
                dos = new DataOutputStream(c.s.getOutputStream());
                dos.writeUTF( Command.ADD+":"+clients.get(i).name +":"+ clients.get(i).state);
            } catch (IOException e) {
                e.printStackTrace();
            }
        }
    }
}
```

addMeToAllUser()方法通过循环，使用 add 命令将新客户端发送给所有已连接的其他客户端。因为新客户端的状态一定是 ready，所以命令的最后一个参数可以直接写成 ready。

```
private void addMeToAllUser(Client c) {
    DataOutputStream dos=null;
    for(int i=0; i<clients.size(); i++){
        if(clients.get(i) != c){
            try {
                dos = new DataOutputStream(clients.get(i).s.getOutputStream());
                dos.writeUTF( Command.ADD + ":" + c.name + ":ready");
            } catch (IOException e) {
                e.printStackTrace();
            }
        }
    }
}
```

在 startServer()方法中的"tellName(c);"语句的后面添加调用这两个方法的语句。

  addAllUserToMe(c);
  addMeToAllUser(c);

3. 创建客户端接收消息的类

因为客户端要不停地接收服务器的各种消息，因此增加一个专门负责客户端接收和发送消息的类 Communication，将发送和接收消息的功能都放在这个类中处理。首先将 FiveClient 类中的连接服务器的功能放在这个类中。

```java
public class Communication {
    FiveClient fc;
    Socket s;
    private DataInputStream dis;
    private DataOutputStream dos;
    public Communication(FiveClient fc) {
        this.fc = fc;
    }
    public void connect(String IP, int port) {
        try {
            s = new Socket(IP,port);
            dis = new DataInputStream(s.getInputStream());
            dos = new DataOutputStream(s.getOutputStream());
        } catch (UnknownHostException e) {
            e.printStackTrace();
        } catch (IOException e) {
            e.printStackTrace();
        }
    }
}
```

因为在后面的程序中，要随时接收和发送消息，因此在 Communication 类中定义数据输入流和数据输出流对象，并在连接后，将 dis 和 dos 对象创建出来。另外类中还有一个 FiveClient 类型的属性和一个 Socket 类型的属性。

在 FiveClient 中添加 Communication 类的属性，添加 connect()方法，并修改监听器类。

添加 Communication 类的属性如下：

Communication c = null;

修改后的监听器类的代码如下：

```java
class ActionMonitor implements ActionListener{
    public void actionPerformed(ActionEvent e) {
        if(e.getSource()==control.exitGameButton){
            System.exit(0);
        }
        else if(e.getSource() == control.connectButton){
            connect();
        }
        else if(e.getSource() == control.cancelGameButton){
            ;
        }
        else if(e.getSource() == control.joinGameButton){
            ;
        }
    }
}
```

在客户端类中添加的 connect()方法如下:
```java
public void connect(){
    c = new Communication(this);
    String ip = control.inputIP.getText();
    c.connect(ip, FiveServer.TCP_PORT);
    message.mesageArea.append("已连接"+"\n");
    isConnected = true;
    control.exitGameButton.setEnabled(true);
    control.connectButton.setEnabled(false);
    control.joinGameButton.setEnabled(true);
    control.cancelGameButton.setEnabled(false);
}
```
在 connect()方法中,创建 Communication 类的对象 c,然后调用 Communication 类的 connect() 方法连接服务器。

4. 客户端接收消息并处理

由于客户端要不断地接收服务器发来的命令,因此为 Communication 类添加内部线程类 ReceiveThread,用于接收服务器的消息。

```java
class ReceiveThread extends Thread{
    Socket s;
    private DataInputStream dis;
    private DataOutputStream dos;
    String msg;
    public ReceiveThread(Socket s){
        this.s = s;
    }
    public void run(){
        while(true){
            try {
                dis = new DataInputStream(s.getInputStream());
                dos = new DataOutputStream(s.getOutputStream());
                msg = dis.readUTF();
                String[] words = msg.split(":");
                if(words[0].equals(Command.TELLNAME)){
                    fc.myname = words[1];
                    fc.userList.userList.add(fc.myname + ":ready");
                    fc.timing.setMyName(fc.myname);
                    fc.message.mesageArea.append("My name: " + fc.myname + "\n");
                }
                else if(words[0].equals(Command.ADD)){
                    fc.userList.userList.add(words[1]+ ":" + words[2]);
                    fc.message.mesageArea.append(words[1]+ ":" + words[2] + "\n");
                }
            }
            catch (IOException e) {
                e.printStackTrace();
```

```
                    return;
                }
            }
        }
    }
```

在 run()方法中不停地从服务器读取消息，根据不同的命令进行不同的处理，目前只处理 tellName 和 add 两种命令。

如果读到的是 tellName 命令，则将读到的用户名分别显示在用户列表区、消息面板区和计时区。

如果读到的是 add 命令，则将读到的用户名和状态添加到用户列表和信息面板中。

最后，在 Communication 类的 connect 方法中创建线程并启动，在"s = **new** Socket(IP,port);"的下一行加入如下代码：

```
new ReceiveThread(s).start();
```

首先运行服务器程序，再运行两个客户端程序，然后分别单击两个客户端的"连接主机"按钮，可以在用户列表区和信息区看到自己和对方的信息。

2.5 实现"加入游戏"按钮的功能

连接服务器之后，用户即可以在用户列表中选择一个对手，然后单击"加入游戏"按钮，申请与该用户下棋，对手可以同意或者拒绝申请。

拒绝申请的处理流程如图 2.11 所示。

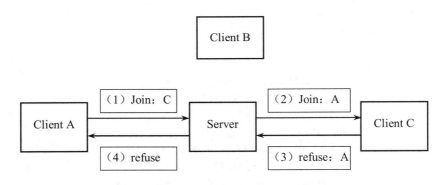

图 2.11 "加入游戏"被拒绝的流程

假设客户端 A 向客户端 C 申请下棋，第一步是客户端 A 向服务器发送 join 命令；第二步服务器向客户端 C 发送 join 命令，通知客户端 A 向其申请下棋；第三步客户端 C 选择拒绝申请，向服务器发送 refuse 命令，表明拒绝；第四步服务器向客户端 A 发送 refuse 命令，通知客户端 C 拒绝了他的申请。

在整个过程中只涉及 A 和 C 两个客户端，与其他客户端无关。

同意申请的处理流程如图 2.12 所示。

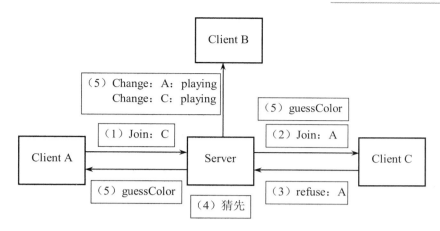

图 2.12 "加入游戏"同意的流程

还是假设客户端 A 向客户端 C 申请下棋，前两步与拒绝的情况相同，第三步客户端 C 选择同意申请，向服务器发送 agree 命令，表明同意与客户端 A 下棋；第四步服务器负责猜先，随机决定两个客户端的黑白子；第五步分别向客户端 A 和客户端 C 发送 guessColor 命令，通知猜先结果，同时向所有客户端发送 change 命令，将客户端用户列表中的客户端 A 和客户端 C 的状态改变为 playing。

2.5.1 客户端申请加入后对方选择同意或拒绝

1. 在 Command 类中增加命令常量

在 Command 类中增加以下五个常量：

> **public static final** String *JOIN* = "join";
> **public static final** String *REFUSE* = "refuse";
> **public static final** String *AGREE* = "agree";
> **public static final** String *CHANGE* = "change";
> **public static final** String *GUESSCOLOR* = "guessColor";

2. 在客户端加入"加入游戏"按钮处理

（1）"加入游戏"按钮监听器。

在 FiveClient 类中，修改其内部监听器类 ActionMonitor 的 actionPerformed()方法，增加"加入游戏"按钮的处理分支。代码如下：

```
public void actionPerformed(ActionEvent e) {
    ……
    else if(e.getSource() == control.joinGameButton){
        String select = userList.userList.getSelectedItem();
        if(select == null){
            message.mesageArea.append("请选择一个对手"+"\n");
            return;
        }
        if(!select.endsWith("ready")){
            message.mesageArea.append("请选择 Ready 状态的对手"+"\n");
            return;
        }
```

```
            if(select.startsWith(myname)){
                message.mesageArea.append("不能选自己作为对手"+"\n");
                return;
            }
            int index = select.lastIndexOf(":");
            String name = select.substring(0,index);
            join(name);
        }
        …
    }
```

用户首先在用户列表区选择一个用户，选择的用户必须是状态为 ready 的客户端，同时不能选择自己，然后单击"加入游戏"按钮。调用 join()方法（在下面介绍），并将对手的用户名作为参数。

在 FiveClient 类中添加方法 join()，代码如下：

```
    public void join(String opponentName){
        c.join(opponentName);
    }
```

在该方法中调用 Communication 类中的 join()方法向服务器发送消息。

（2）向服务器发送消息。

在 Communication 类中添加 join()方法，代码如下：

```
    public void join(String opponentName) {
        try {
            dos.writeUTF(Command.JOIN + ":" + opponentName);
        } catch (IOException e) {
            e.printStackTrace();
        }
    }
```

方法中向服务器发送 join 命令，join 命令的格式是"join:对手用户名"。

3. 服务器接收 join 命令并发送给对手

由于服务器要不停地从客户端接收消息，因此在服务器端也要创建一个内部线程类 ClientThread，用于处理从客户端接收到的消息，并在 startServer()方法中启动线程。

内部线程类 ClientThread 的代码如下：

```
    class ClientThread extends Thread{
        private Client c;
        private DataInputStream dis;
        private DataOutputStream dos;
        ClientThread(Client c){
            this.c = c;
        }
        public void run(){
            while(true){
                try {
                    dis = new DataInputStream(c.s.getInputStream());
                    String msg = dis.readUTF();
                    String[] words = msg.split(":");
```

```
                if(words[0].equals(Command.JOIN)){
                    String opponentName = words[1];
                    for(int i=0; i<clients.size(); i++){
                        if(clients.get(i).name.equals(opponentName)){
                            dos = new DataOutputStream(clients.get(i).s.getOutputStream());
                            dos.writeUTF(Command.JOIN + ":" + c.name);
                            break;
                        }
                    }
                }
            }
            catch (IOException e) {
                e.printStackTrace();
                return;
            }
        }
    }
}
```

然后在 startServer()方法的最后一行"addAllUserToMe(c);"的后面加入以下语句来启动线程：

new ClientThread(c).start();

每当一个客户端连接到服务器上，就为该客户端创建一个线程类对象并启动线程，用来接收该客户端发来的消息。收到消息后，根据不同的命令进行不同的处理。目前只有一个 join 命令，服务器收到 join 命令后，在客户端链表中找到对手客户端，然后向其发送 join 命令，参数为申请客户端的用户名。

4. 对手接到命令将选择结果发送给服务器

当客户端收到下棋邀请时，出现一个对话框，让用户选择同意还是拒绝邀请，如图 2.13 所示，该图是 Player1 向 Player2 发出邀请后在 Player2 端显示的界面。

图 2.13　收到邀请时的对话框

用户单击"接受"或"拒绝"按钮后，将选择结果发给服务器，如果用户在规定的时间内没有选择，则默认为拒绝，因此该对话框应该具有计时功能。下面创建一个带有计时功能的对话框。

（1）创建计时对话框类。

创建计时对话框类 TimeDialog，代码如下：

```java
public class TimeDialog {
    private String message = null;
    private int seconds = 0;
    private JLabel label = new JLabel();
    private JButton confirm;
    private JButton cancel;
    private JDialog dialog = null;
    int result = -5;
    public int   showDialog(Frame father, String message, int sec) {
        this.message = message;
        seconds = sec;
        label.setText(message);
        label.setBounds(80,6,200,20);
        ScheduledExecutorService s = Executors.newSingleThreadScheduledExecutor();
        confirm = new JButton("接受");
        confirm.setBounds(100,40,60,20);
        confirm.addActionListener(new ActionListener() {
            public void actionPerformed(ActionEvent e) {
                result = 0;
                TimeDialog.this.dialog.dispose();
            }
        });
        cancel = new JButton("拒绝");
        cancel.setBounds(190,40,60,20);
        cancel.addActionListener(new ActionListener() {
            public void actionPerformed(ActionEvent e) {
                result = 1;
                TimeDialog.this.dialog.dispose();
            }
        });
        dialog = new JDialog(father, true);
        dialog.setTitle("提示：本窗口将在"+seconds+"秒后自动关闭");
        dialog.setLayout(null);
        dialog.add(label);
        dialog.add(confirm);
        dialog.add(cancel);
        s.scheduleAtFixedRate(new Runnable() {
```

```
            public void run() {
                    TimeDialog.this.seconds--;
                    if(TimeDialog.this.seconds == 0) {
                            TimeDialog.this.dialog.dispose();
                    }else {
                            dialog.setTitle("本窗口将在"+seconds+"秒后关闭");
                    }
            }
    }, 1, 1, TimeUnit.SECONDS);
    dialog.pack();
    dialog.setSize(new Dimension(350,100));
    dialog.setLocationRelativeTo(father);
    dialog.setVisible(true);
    return result;
}
```

这个对话框除了使用 ScheduledExecutorService 接口之外，并没有什么特别之处。ScheduledExecutorService 提供了按时间安排执行任务的功能，它的 scheduleAtFixedRate()方法安排所提交的 Runnable 任务按指定的间隔重复执行，方法的原型如下：

   scheduledfuture schedulewithfixeddelay(runnable command, long initialdelay,long delay,timeunit unit)

参数的含义如下：

command：执行的任务（线程对象）。

initialdelay：首次执行的延迟时间。

delay：一次执行终止和下一次执行开始之间的延迟。

unit：initialdelay 和 delay 参数的时间单位。

在我们的程序中时间单位设置为秒（TimeUnit.SECONDS），两个时间间隔都是 1，也就是说每隔一秒执行一次，而执行的任务就是第一个参数，每次将对话框的显示时间减 1，如果这个时间变为 0，则让对话框消失，否则将这个剩余时间显示在对话框的标题中。

对话框的初始显示时间是在 showDialog()方法中通过参数获得的，因此在显示对话框时，要指定这个显示时间。

对话框还有一个属性 result，也是 showDialog()方法的返回值，将其初始化为-5，如果单击"接受"按钮，将 result 赋值为 0；如果单击"拒绝"按钮，将 result 赋值为 1。因此可以通过 showDialog()方法的返回值判断用户是否选择了"接受"按钮，如果返回值为 0，则单击了"接受"按钮，否则就是单击了"拒绝"按钮或超时没有处理。

（2）修改 Communication 类。

在 Communication 类中修改其内部类 ReceiveThread 中的 run 方法，添加 join 处理的分支。

```
            else if(words[0].equals(Command.JOIN)){
                    String name = words[1];
                    TimeDialog d = new TimeDialog();
                    int select = d.showDialog(fc, name + " 邀请你下棋，是否接受？", 100);
                    if(select == 0){
```

```
                dos.writeUTF(Command.AGREE + ":" + name);
            }
            else{
                dos.writeUTF(Command.REFUSE + ":" + name);
            }
        }
```

创建 TimeDialog 类的对象并显示该对话框。在显示对话框时，通过参数指定主窗口为对话框的父窗口，以及对话框要显示的信息和显示的时间，根据对话框的返回值确定向服务器发送 agree 命令或 refuse 命令。

2.5.2 完成猜棋并准备好下棋

1. 服务器收到拒绝或同意的处理

（1）拒绝的处理。

如果服务器收到客户端拒绝邀请的命令，则直接向申请客户端发送拒绝命令即可。在服务器端接收消息的线程中添加处理 refuse 命令的代码如下：

```
        else if(words[0].equals(Command.REFUSE)){
            String opponentName = words[1];
            for(int i=0; i<clients.size(); i++){
                if(clients.get(i).name.equals(opponentName)){
                    dos = new DataOutputStream(clients.get(i).s.getOutputStream());
                    dos.writeUTF(Command.REFUSE + ":" + c.name);
                    break;
                }
            }
        }
```

（2）同意的处理。

如果服务器收到客户端同意邀请的命令，则改变两个用户为 playing 状态，然后猜先，并将猜先结果通知两个客户端。

在 Command 类中添加 GUESSCOLOR 常量。

```
        public static final String GUESSCOLOR = "guessColor";
```

在服务器接收消息的线程中添加处理 agree 的代码如下：

```
        else if(words[0].equals(Command.AGREE)){
            c.state = "playing";
            String opponentName = words[1];
            //两个客户相互为对手
            for(int i=0; i<clients.size(); i++){
                if(clients.get(i).name.equals(opponentName)){
                    clients.get(i).state = "playing";
                    clients.get(i).opponent = c;
                    c.opponent = clients.get(i);
                    break;
                }
```

```
            }
        //改变所有客户端中这两个客户的状态为 playing
        for(int i=0; i<clients.size(); i++){
                dos = new dataoutputstream(clients.get(i).s.getoutputstream());
                dos.writeutf(command.change + ":" + c.name + ":playing");
                dos.writeutf(command.change + ":" +opponentname + ":playing");
        }
        int r = (int) (math.random()*2);    //随机分配黑棋、白棋
        if(r==0){
                dos = new dataoutputstream(c.s.getoutputstream());
                dos.writeutf(command.guesscolor+ ":black:" + opponentname);
                dos = new dataoutputstream(c.opponent.s.getoutputstream());
                dos.writeutf(command.guesscolor+ ":white:" + c.name);
        }
        else
        {
                dos = new dataoutputstream(c.s.getoutputstream());
                dos.writeutf(command.guesscolor+ ":white:" + opponentname);
                dos = new dataoutputstream(c.opponent.s.getoutputstream());
                dos.writeutf(command.guesscolor+ ":black:" + c.name);
        }
        tamessage.append(c.name + " playing\n");
        tamessage.append(opponentname + " playing\n");
    }
```

首先将服务器中的客户端链表中的两个客户端设置成互为对手，并将状态改为 playing；然后向所有客户端发送两个 change 命令，将这两个下棋客户端的状态改为 playing；最后随机猜先，分别向这两个客户端发送 guessColor 命令，通知它们猜到的是黑棋或者白棋。

2. 客户端收到拒绝或同意的处理

如果客户端收到服务器发来的拒绝命令，就出现一个信息框，显示对方拒绝的信息；如果收到猜先的命令，则根据是黑棋还是白棋，进行相应的处理，准备好下棋。

（1）PanelBoard 类加入属性。

在 PanelBoard 类中加入属性 isGoing，表示是否轮到自己下棋。true 代表轮到自己下棋，false 代表轮到对手下棋。

```
        boolean isGoing = false;
```

（2）Communication 类处理同意或拒绝的消息。

在 Communication 类的内部类 ReceiveThread 中，为 run()方法中增加处理被拒绝的消息、猜棋的消息和改变用户状态的消息分支，代码如下：

```
        else if(words[0].equals(Command.REFUSE)){
                String name = words[1];
                JOptionPane.showMessageDialog(fc, name + " 拒绝了您的邀请！");
        }
        else if(words[0].equals(Command.CHANGE)){
```

```java
            String name = words[1];
            String state = words[2];
            for(int i=0; i<fc.userList.userList.getItemCount(); i++){
                if(fc.userList.userList.getItem(i).startsWith(name)){
                    fc.userList.userList.replaceItem(name + ":" +state, i);
                }
            }
            fc.message.mesageArea.append(name + " " + state + "\n");
        }
        else if(words[0].equals(Command.GUESSCOLOR)){
            String color = words[1];
            String oppName = words[2];
            fc.board.isGamming = true;
            fc.opname = oppName;
            fc.timing.setOpName(oppName);
            if(color.equals("black")){        //黑棋
                fc.timing.setMyIcon("black");
                fc.timing.setOpIcon("white");
                fc.board.isBlack = true;
                fc.board.isGoing = true;
            }
            else if(color.equals("white")){ //白棋
                fc.timing.setMyIcon("white");
                fc.timing.setOpIcon("black");
                fc.board.isBlack = false;
                fc.board.isGoing = false;
            }
            fc.control.joinGameButton.setEnabled(false);
            fc.control.cancelGameButton.setEnabled(true);
            fc.control.exitGameButton.setEnabled(false);
            fc.message.mesageArea.append("My color is " + color + "\n");
        }
```

拒绝消息的处理比较简单，只需要出现一个信息框，显示被对手拒绝的信息就可以了。

改变用户的状态，首先在用户列表中根据用户名找到指定的用户，然后将该用户的状态改变成指定的状态。

如果收到猜棋的消息，则首先将棋盘类 isGamming 设置为 true，表示已经处于下棋状态，并将对手的名字赋给 FiveClient 类中的属性 opname，同时将对手的用户名显示在计时面板中。

然后根据猜棋结果进行一些设置，比如猜到黑棋，就要将自己的头像设置为黑色图像、将对手的头像设置为白色图像、将棋盘类的 isBlack 属性和 isGoing 属性都设置为 true。如果猜到白棋则正好相反。

最后设置控制面板按钮的状态，开始下棋后，"加入游戏"按钮和"退出程序"按钮将被禁用。

（3）修改 PanelBoard 类的内部类 MouseMonitor。

修改 MouseMonitor 类的 mousePressed()方法，修改后的代码如下：

```
public void mousePressed(MouseEvent e){//鼠标在组件上按下时调用
    if(!isGamming) return;
    if(!isGoing) return;         //不是自己下棋
    //将鼠标点击的坐标位置转换成网格索引
    int col=(e.getX()-MARGIN+SPAN/2)/SPAN;
    int row=(e.getY()-MARGIN+SPAN/2)/SPAN;
    //落在棋盘外不能下
    if(row<0||row>ROWS||col<0||col>COLS)
        return;
    //如果 x、y 位置已经有棋子存在，不能下
    if(hasChess(col, row)){
        return;
    }
    Chess ch=new Chess(PanelBoard.this,col, row, isBlack?Color.black:Color.white);
    chessList[chessCount++]=ch;
    repaint();           //通知系统重新绘制
    isGoing = false;
    //向服务器发送下棋信息（稍后实现）
    if(isWin(col, row)){
        //如果胜出则向服务器发送信息（稍后实现）
    }
}
```

这个方法与单机版差不多，不同的地方有以下几点：一是增加一个能否下棋的判断条件，如果未轮到自己下棋，则直接返回；二是下完一个棋子后，不能像单机版那样转换棋子的颜色，而是将 isGoing 设置为 false，表示轮到对手下棋，并将下棋的信息发送给服务器；三是如果赢棋，要向服务器发送赢棋消息，而不是像单机版那样显示一个赢棋的信息，这一处理在稍后介绍。

启动服务器后，运行两个客户端，连接主机，申请下棋，对方同意申请后，目前只能由黑方下一个棋子，因为还没有将黑子的信息通过服务器发送给对手，我们将在 2.6 节实现下棋功能。

2.6　实现下棋功能

假设客户端 A 与客户端 C 下棋，下棋过程的流程如图 2.14 所示。

图 2.14 的上方显示，客户端 A 下棋后，通过 go 命令将下棋信息发送给服务器，服务器再转发给客户端 C，这一过程与其他客户端无关。

图 2.14 的下方显示的是赢棋后的流程，客户端 A 赢棋后，向服务器发送 win 命令，然后服务器向所有客户端发送 change 命令，修改这两个客户端的状态为 ready，再向客户端 A 和客户端 C 发送 tellResult 命令，通知下棋结果。最后客户端做结束下棋的处理。

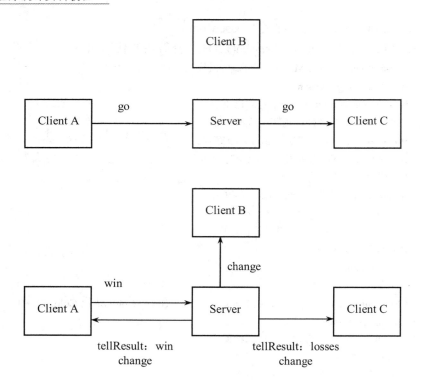

图 2.14　下棋过程的流程

2.6.1　客户端向服务器发送下棋消息

1. 在 Command 类中添加命令

在 Command 类中添加如下几个命令：
  public static final String *GO* = "go";
  public static final String *WIN* = "win";
  public static final String *TELLRESULT* = "tellResult";

2. 在 Communication 类中添加方法

在 Communication 类中添加向服务器发送 go 命令和 wins 命令的方法。

go 命令有两个参数，即棋子的列坐标和行坐标，格式为"go:col:row"，代码如下：

```
public void go(int col, int row) {
    try {
        String msg = Command.GO+ ":" + col + ":" + row;
        dos.writeUTF(msg);
    } catch (IOException e) {
        e.printStackTrace();
    }
}
```

win 命令没有参数，代码如下：

```
public void wins() {
    try {
```

```
                dos.writeUTF(Command.WIN);
            } catch (IOException e) {
                e.printStackTrace();
            }
        }
```

3. 在 PanelBoard 类中调用上面两个方法

（1）在 PanelBoard 类中添加对 FiveClient 类的引用。

PanelBoard 类不能直接调用 Communication 类的方法，只能通过 FiveClient 类调用，因此要在 PanelBoard 类中添加一个 FiveClient 类的引用，代码如下：

```
        FiveClient fc;
```

修改 PanelBoard 类的构造方法，增加一个 FiveClient 类的参数，通过参数为 fc 初始化。

```
        public PanelBoard(FiveClient fc){
            this.fc = fc;
            img=Toolkit.getDefaultToolkit().getImage("img/board.jpg");
            this.addMouseListener(new MouseMonitor());
            this.addMouseMotionListener(new MouseMotionMonitor());
        }
```

修改 FiveClient 类的构造方法，在创建棋盘类对象时传递参数。

```
        board=new PanelBoard(this);
```

（2）在 PanelBoard 类中调用 go()方法和 wins()方法。

找到 PanelBoard 类的内部类 MouseMonitor，将 mousePressed()方法中最后部分改成下面的形式。

```
        isGoing = false;
        fc.c.go(col,row);
        if(isWin(col, row)){
            fc.c.wins();
        }
```

每下一手棋，都要通过调用 Communication 类的 go()方法将棋子信息发送给服务器，然后判断是否赢棋，如果赢棋，则通过调用 Communication 类的 wins()方法将赢棋信息发送给服务器。

2.6.2 服务器接收消息并处理

服务器如果收到 go 命令，可以直接将命令消息转发给对手；如果收到 win 命令，则要向所有客户端发送两个 change 命令，将这两个客户端的状态改为 ready，然后分别向两个客户端发送 tellResult 命令，通知赢棋和输棋。

在服务器接收消息的线程中添加处理 go 命令和 win 命令的代码如下：

```
        else if(words[0].equals(Command.GO)){
            dos = new DataOutputStream(c.opponent.s.getOutputStream());
            dos.writeUTF(msg);
            taMessage.append(c.name + " " +msg + "\n");
        }
        else if(words[0].equals(Command.WIN)){
```

```java
            //在所有客户端的客户列表中,将这两个客户的状态改为ready
            for(int i=0; i<clients.size(); i++){
                dos = new DataOutputStream(clients.get(i).s.getOutputStream());
                dos.writeUTF(Command.CHANGE+ ":" + c.name + ":ready");
                dos.writeUTF(Command.CHANGE+ ":" + c.opponent.name + ":ready");
            }
            dos = new DataOutputStream(c.s.getOutputStream());
            dos.writeUTF(Command.TELLRESULT+ ":win");       //向自己发回胜利命令
            dos = new DataOutputStream(c.opponent.s.getOutputStream());
            dos.writeUTF(Command.TELLRESULT+ ":losses");    //向对方发送失败命令
            c.state = "ready";
            c.opponent.state = "ready";
            taMessage.append(c.name + " win\n");
            taMessage.append(c.opponent.name + " loss\n");
        }
```

2.6.3 客户端接收消息并处理

1. Communication 类接收消息

在接收消息的线程类中,增加处理 go 命令和 tellResult 命令的分支。处理 go 命令的分支如下:

```java
else if(words[0].equals(Command.GO)){
    int col = Integer.parseInt(words[1]);
    int row = Integer.parseInt(words[2]);
    fc.board.addOpponentChess(col,row);
}
```

首先获取对手下棋的列坐标和行坐标,然后调用棋盘类的 addOpponentChess()方法将对手下的棋子显示在棋盘中(addOpponentChess()方法稍后介绍)。

处理 tellResult 命令的分支如下:

```java
else if(words[0].equals(Command.TELLRESULT)){
    if(words[1].equals("win"))
        fc.board.winsGame();
    else
        fc.board.lossesGame();
    fc.control.joinGameButton.setEnabled(true);
    fc.control.cancelGameButton.setEnabled(false);
    fc.control.exitGameButton.setEnabled(true);
}
```

如果赢棋,调用棋盘类的 winsGame()方法进行赢棋处理(winsGame()方法稍后介绍);如果输棋,调用棋盘类的 lossesGame()方法进行输棋处理(lossesGame ()方法稍后介绍),最后设置按钮的状态。

2. PanelBoard 类添加方法

在 PanelBoard 类中添加四个方法,分别是添加对手的棋子、赢棋处理、输棋处理和棋局结束后的处理。

(1) 添加对手棋子的方法。

添加对手棋子的方法如下：

```
public void addOpponentChess(int col, int row) {
    Chess ch=new Chess(this, col, row, isBlack? Color.white:Color.black);
    chessList[chessCount++]=ch;
    isGoing = true;
    repaint();//通知系统重新绘制
}
```

方法中首先创建对手的棋子，棋子的颜色正好和自己棋子的颜色相反，然后将棋子加入到棋子数组中，将 isGoing 设置为 true，使自己处于下棋状态。

(2) 赢棋处理的方法。

赢棋处理的方法如下：

```
public void winsGame() {
    resetGame();
    String colorName=isBlack?"黑棋":"白棋";
    String msg=String.format("恭喜，%s 赢了！", colorName);
    JOptionPane.showMessageDialog(PanelBoard.this, msg);
}
```

首先调用方法 resetGame() 将棋盘的一些参数重新初始化（resetGame() 方法稍后介绍），以便之后重新申请下棋，然后显示赢棋信息。

(3) 输棋处理的方法。

输棋处理的方法如下：

```
public void lossesGame() {
    resetGame();
    String colorName=isBlack?"黑棋":"白棋";
    String msg=String.format("遗憾，%s 输了！", colorName);
    JOptionPane.showMessageDialog(PanelBoard.this, msg);
}
```

首先调用方法 resetGame() 将棋盘的一些参数重新初始化，然后显示输棋信息。

(4) 棋局结束的处理方法。

棋局结束的处理方法如下：

```
public void resetGame(){
    chessCount =0; //当前棋盘棋子个数
    isGamming = false;
    //清除棋子
    for(int i=0;i<chessList.length;i++){
        chessList[i]=null;
    }
    repaint();
    fc.control.joinGameButton.setEnabled(true);
}
```

在 resetGame() 方法中，将棋盘类中的部分变量重新赋值，结束下棋状态，并准备重新下棋。

至此，下棋功能已经全部实现，在后面的几节中将实现下棋的计时功能、"放弃游戏"功能，以及"关闭程序"的额外处理。

2.7 实现"放弃游戏"按钮的功能

在下棋过程中，用户可以单击"放弃游戏"按钮认输，结束下棋。单击"放弃游戏"按钮后的流程如图 2.15 所示。

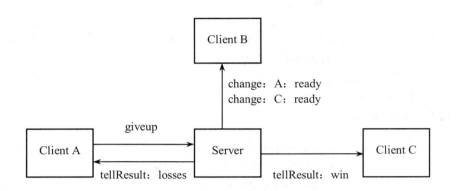

图 2.15 放弃游戏的流程

例如客户端 A 与客户端 C 下棋，客户端 A 单击"放弃游戏"按钮后，向服务器发送 giveup 命令认输，服务器收到 giveup 命令后，向所有客户端发送 2 次 change 命令，将客户端 A 和客户端 C 的状态改为 ready，并向客户端 A 和客户端 C 分别发送 tellResult 命令，通知输棋和赢棋。因为在前面的程序中，客户端已经对赢棋和输棋的命令进行了处理，因此，在下面的程序中只需要服务器端发送 tellResult 命令，而客户端的响应使用前面相同的程序即可。

2.7.1 Command 类添加常量

在 Command 类中添加下面的常量：
```
public static final String GIVEUP = "giveup";
```

2.7.2 添加"放弃游戏"按钮的响应代码

在 FiveClient 的内部类 ActionMonitor 中找到"放弃游戏"的处理分支，并加入下面的代码：
```
else if(e.getSource() == control.cancelGameButton){
    c.giveup();
}
```

2.7.3 在 Communication 类中添加 giveup()方法

在 Communication 类中添加 giveup()方法，向服务器发送 giveup 命令，giveup()方法的代码如下：
```
public void giveup() {
    try {
        dos.writeUTF(Command.GIVEUP);
    } catch (IOException e) {
```

2.7.4 服务器接收 giveup 命令并处理

找到 FiveServer 类的内部类 ClientThread，在它的 run()方法中添加处理 giveup 命令的代码。

```
        else if(words[0].equals(Command.GIVEUP)){
            for(int i=0; i<clients.size(); i++){
                dos = new DataOutputStream(clients.get(i).s.getOutputStream());
                dos.writeUTF(Command.CHANGE+ ":" + c.name + ":ready");
                dos.writeUTF(Command.CHANGE+ ":" + c.opponent.name + ":ready");
            }
            dos = new DataOutputStream(c.s.getOutputStream());
            dos.writeUTF(Command.TELLRESULT+ ":losses");
            dos = new DataOutputStream(c.opponent.s.getOutputStream());
            dos.writeUTF(Command.TELLRESULT+ ":win");
            c.state = "ready";
            c.opponent.state = "ready";
            taMessage.append(c.name + " loss\n");
            taMessage.append(c.opponent.name + " win\n");
        }
```

服务器收到 giveup 命令后，向所有客户端发送 change 命令，将两个客户端的状态设置为 ready，然后向这两个客户端分别发送 tellResult 命令，通知赢棋或输棋，同时将这两个客户端的状态也设置为 ready。

2.8 加入计时功能

每局棋规定每位棋手下棋的总时间，时间用完自动判负。

2.8.1 设计计时线程类

设计用于倒计时的线程类 TimerThread，代码如下：

```
public class TimerThread extends Thread {
    FiveClient fc;
    private int myTotalTime;
    private int opTotalTime;
    public TimerThread(FiveClient fc, int totalTime) {
        this.fc = fc;
        this.myTotalTime = totalTime;
        this.opTotalTime = totalTime;
    }
    public void run(){
        fc.timing.setMyTime(myTotalTime);
```

```
                fc.timing.setOpTime(opTotalTime);
                while(fc.board.isGamming){
                    try {
                        sleep(1000);
                    } catch (InterruptedException e) {
                        e.printStackTrace();
                    }
                    if(fc.board.isGoing){
                        myTotalTime--;
                        fc.timing.setMyTime(myTotalTime);
                        if(myTotalTime<=0){
                            fc.c.giveup();
                            break;
                        }
                    }
                    else{
                        opTotalTime--;
                        fc.timing.setOpTime(opTotalTime);
                        if(opTotalTime<=0)
                            break;
                    }
                }
            }
```

TimerThread 类在构造方法中设置棋手的总用时，启动线程后，执行 run()方法，将总的用时显示在计时区。在循环中（循环条件是正在下棋），每隔一秒计时一次，如果是自己在下棋，则自己的总时间减掉 1 秒，重新在计时区显示自己的剩余时间；然后判断剩余时间是否小于等于 0，是的话，调用 Communication 类中的 giveup()方法认输；再执行 break 语句结束循环，线程终止。如果是对手下棋，则将对手的剩余时间减掉 1 秒，再将对手的剩余时间显示在计时区。每个客户端只负责将自己的赢棋或输棋的信息发送给服务器，因此即使对手的剩余时间为 0，也不能替对手发送，只能等待对手发送，但线程不能再继续计时，所以用 break 语句结束循环，从而结束线程。

2.8.2 猜先后启动倒计时线程

在 Communication 类添加 TimerThread 类的属性如下：
```
TimerThread tt;
```
找到 Communication 类的内部类 ReceiveThread，在处理 guessColor 命令的分支最后加上下面的两行代码，启动计时线程。
```
tt = new TimerThread(fc,120);
tt.start();
```
在创建线程类对象时，将每方用时设置为 2 分钟（120 秒）。

2.9 完善"关闭程序"按钮的功能

当一个客户端关闭程序时,需要通知其他客户端,将该客户端从用户列表中删除,流程如图 2.16 所示。

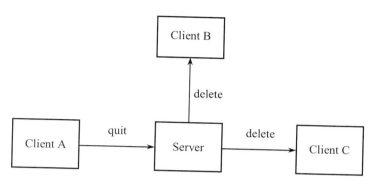

图 2.16 关闭程序的流程

客户端 A 关闭程序时,向服务器发送 quit 命令,然后服务器向其他客户端发送 delete 命令,通知将客户端 A 从其用户列表中删除。

2.9.1 在 Command 类中添加命令

在 Command 类中添加 quit 和 delete 命令常量。
  public static final String *QUIT* = "quit";
  public static final String *DELETE* = "delete";

2.9.2 客户端向服务器发送命令

1. 修改"关闭程序"按钮的响应代码

在 FiveClient 的内部类 ActionMonitor 中找到"关闭程序"的处理分支,进行如下修改:
```
if(e.getSource()==control.exitGameButton){
    if(isConnected){
        c.disConnect();
    }
    System.exit(0);
}
```
如果用户已经连接到服务器,则需要调用 Communication 类的 disConnect()方法通知服务器;如果还没有与服务器连接,当然不需要通知服务器,直接退出程序。

2. 在 Communication 类中添加方法

在 Communication 类中添加 disConnect()方法,向服务器发送 quit 命令,代码如下:
```
public void disConnect() {
    try {
        dos.writeUTF(Command.QUIT);
    } catch (IOException e) {
```

2.9.3 服务器处理 quit 命令

找到 FiveServer 类的内部类 ClientThread，在它的 run()方法中添加处理 quit 命令的分支，代码如下：

```
else if(words[0].equals(Command.QUIT)){
    for(int i=0; i<clients.size(); i++){
        if(clients.get(i)!=c){
            dos = new DataOutputStream(clients.get(i).s.getOutputStream());
            dos.writeUTF(Command.DELETE+":" + c.name);
        }
    }
    clients.remove(c);
    taMessage.append(c.name   + " quit\n");
    clientNum--;
    lStatus.setText("连接数" + clientNum);
    return;
}
```

首先向其他所有客户端发送 delete 命令，通知客户端将该用户从用户列表中删除；然后将该用户从服务器的客户端链表中删除，并将连接的客户端数 clientNum 减 1；最后通过 return 命令结束线程。

2.9.4 客户端处理 delete 命令

在 Communication 类中添加处理 delete 命令，代码如下：

```
else if(words[0].equals(Command.DELETE)){
    for(int i=0; i<fc.userList.userList.getItemCount();i++){
        String name = fc.userList.userList.getItem(i);
        if(name.startsWith(words[1])){
            fc.userList.userList.remove(i);
        }
    }
    fc.message.mesageArea.append(words[1] + " disconnected" + "\n");
}
```

客户端收到 delete 命令后，将该用户从用户列表中删除。

至此，网络五子棋的基本功能已经完成。下一章将实现用户注册、用户登录、记录棋谱数据、棋谱回放等功能。

2.10 作业

1. 修改单机版五子棋程序，在窗口的下方增加一个组件，用于提示该哪一方下棋，并且在下棋结束后显示祝贺赢棋的信息，如图 2.17 所示。

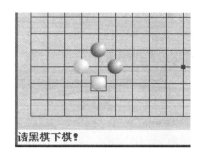 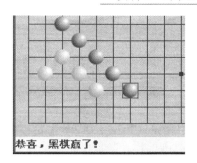

图 2.17 作业 1 图

2．在下棋时，有时会临时改变主意，比如按下鼠标的瞬间突然不想在这个位置下子了，这时希望能够在不松开鼠标的情况下将鼠标移到棋盘的外面再松开，不完成下棋的动作，或者将鼠标移动到另外一个位置松开，将棋子下在松开的位置，目前的程序是不支持这一功能的，如何修改程序实现这一功能。

3．在作业 2 的基础上，实现按下鼠标拖动时，能够看到棋子在棋盘上的拖动情况，松开鼠标时，在松开的位置下棋。

4．目前的程序每方的总用时是在程序中指定的，如果用户希望在申请对局时自己指定用时时间，程序中应如何处理？

5．在我们的程序中，对于异常只是调用 printStackTrace() 方法输出异常信息，并没有进行实质上的处理，这显然是不够的。比如在服务器还没有启动时，当客户端程序启动后，单击"连接服务器"按钮，就会产生异常，客户端程序应该显示未连接成功的提示信息，这时不能进行已连接成功的相关处理。应该如何改进程序以较好地处理这个异常问题？

实训 3　棋谱的保存与回放

为了保存下棋的数据，用户必须先注册，登录服务器才能下棋，棋局结束后，将棋局保存在数据库中，将棋谱保存在数据文件中。将来可以根据用户名查询出该用户下过的所有棋局，选择其中的某一个棋局并将棋谱读出来回放。在这里我们使用的数据库是 MySQL 数据库。

3.1　创建数据库

3.1.1　数据库设计

为了保存用户信息和棋局信息，我们先创建一个数据库 fiveChess，数据库中包含两个数据表 user 和 game。

user 表用于保存用户信息，字段包括 id、姓名、密码、电子邮箱、级别和注册日期，表结构如表 3.1 所示。

表 3.1　user 表结构

字段名	类型	长度	允许空	备注
userid	INT		否	主键，自动增长
name	VARCHAR	10	否	用户名
password	VARCHAR	10	否	密码
email	VARCHAR	30	否	邮箱
level	INT		否	级别
regdate	Date		否	注册日期

game 表用于保存棋局信息，字段包括 id、下棋日期、黑方、白方、赢棋方和记录棋谱的文件名，表结构如表 3.2 所示。

表 3.2　game 表结构

字段名	类型	长度	允许空	备注
gameid	INT		否	主键，自动增长
gameDate	Date		否	下棋日期
playerBlack	VARCHAR	10	否	黑方
playerWhite	VARCHAR	10	否	白方
winner	VARCHAR	10	否	赢棋方
manualFileName	VARCHAR	100	否	记录棋谱的文件名

3.1.2 数据库创建

在后面的程序中还会设计很多类，因此我们建立 3 个包，将实现下棋功能的类放在 five 包中，与用户有关的类放在 user 包中，与棋局和棋谱有关的类放在 game 包中。

建立 five 包，将实训 2 程序中的类全部移到 five 包中。

为了在 Java 程序中操作数据库中的数据，需要配置数据库驱动程序 JDBC。如果计算机中没有，可从 MySQL 网站上下载 mysql-connector-java 压缩文件，然后解压缩。

在 Eclipse 中右击工程名 NetFive，在快捷菜单中选择 Build Path→Add External Archives 菜单项，在 JAR Selection 对话框中找到上面解压缩的 jar 文件，如图 3.1 所示，单击"打开"按钮，将该文件导入到我们的工程中。

图 3.1 JAR Selection 对话框

新建 user 包，在 user 包中新建类 CreateDatabase，该类负责创建数据库 fiveChess 以及数据库中的两个表 user 和 game，代码如下：

```
public class CreateDatabase {
    public static void main(String[] args) {
        Connection conn=null;
        Statement stmt = null;
        try {
            Class.forName("com.mysql.jdbc.Driver");
            conn = DriverManager.getConnection("jdbc:mysql://localhost:3306/","root","1234");
            stmt = conn.createStatement();
        }
        catch (ClassNotFoundException e) {
```

```java
                e.printStackTrace();
                System.exit(-1);
            }
        catch (SQLException e) {
            e.printStackTrace();
            try {
                conn.close();
            }
            catch (SQLException e1) {
                e1.printStackTrace();
                System.exit(-1);
            }
        }
        createDatabase(stmt, "fiveChess");
        try {
            stmt.close();
            conn.close();
        }
        catch (SQLException e) {
            e.printStackTrace();
        }
    }
    public static void createDatabase(Statement stmt, String dbName){
        try {
            stmt.executeUpdate("create database " + dbName);
            stmt.executeUpdate("use " + dbName);
            String sql;
            sql = "create table user(id int auto_increment not null primary key," +
                    "name VARCHAR(10) not null, " +
                    "password VARCHAR(10) not null," +
                    "email VARCHAR(30) not null," +
                    "level int not null," +
                    "regDate date not null)";
            stmt.executeUpdate(sql );
            sql = "insert into user(name,password,email,level,regDate) " +
                    "values('test1', 'test1','test1@five.com',1,'2016-01-22')";
            stmt.executeUpdate(sql);
            sql = "insert into user(name,password,email,level,regDate) " +
                    "values('test2', 'test2','test2@five.com',1,'2016-01-22')";
            stmt.executeUpdate(sql);
            sql = "create table game(id int auto_increment not null primary key," +
                    "gameDate date not null, " +
                    "playerBlack VARCHAR(10) not null, " +
                    "playerWhite VARCHAR(10) not null, " +
                    "winner VARCHAR(10) not null, " +
                    "manualFileName VARCHAR(100) not null)";
```

```
                stmt.executeUpdate( sql );
                sql="insert into game(gameDate, playerBlack, playerwhite, winner,manualFileName)"
                    + " values('2016-01-22','test1','test2','test2','game1201601221205.fiv')";
                stmt.executeUpdate(sql);
                sql = "insert into game(gameDate, playerBlack, playerwhite, winner,manualFileName)"
                    + " values('2016-01-22','test1','test2','test1','game1201601221848.fiv')";
                stmt.executeUpdate(sql);
            }
            catch (SQLException e) {
                e.printStackTrace();
            }
        }
    }
```

在主方法中创建数据库连接 Connection 对象和 Statement 对象，然后调用 createDatabase() 方法，创建数据库 fiveChess，再创建 user 表，并向 user 表插入两条记录，最后创建 game 表，并向 game 表插入两条记录。给两个表分别插入两条记录是为了测试一下创建数据库和表的功能是否完善，实际中不必插入这些记录。

运行 CreateDatabase 类，创建数据库和表，以后的程序是在这两个表的基础上实现的。创建好数据库之后，就不要再次运行 CreateDatabase 类了，以后如果需要初始化数据库，可以在这个类的基础上稍加修改。

3.2 用户管理

为了实现用户注册和登录的功能，我们创建用户管理类，实现用户的添加和查找功能。

3.2.1 数据库连接类

由于在后面的程序中经常要进行数据库连接操作，因此我们首先设计一个连接数据库的类 DBConnection，将其放在 user 包中，代码如下：

```
        public class DBConnection {
            Connection conn=null;
            public DBConnection(){
                try {
                    Class.forName("com.mysql.jdbc.Driver");
                    conn = DriverManager.getConnection(
                        "jdbc:mysql://localhost:3306/fiveChess","root","1234");
                }
                catch (ClassNotFoundException e) {
                    e.printStackTrace();
                    System.exit(-1);
                }
                catch (SQLException e) {
                    e.printStackTrace();
                    if(conn!=null){
```

```java
            try {
                conn.close();
            }
            catch (SQLException e1) {
                e1.printStackTrace();
                System.exit(-1);
            }
        }
    }
    public Connection getConn() {
        return conn;
    }
}
```

DBConnection 类的代码比较简单，在构造方法中创建 Connection 对象。其他类可以调用 DBConnection 类的 getConn()方法中获得这个数据库连接对象。

3.2.2 用户管理

1. User 类

在 user 包中添加 User 类，因为在后面的注册程序中，需要在客户端与服务器之间传递 User 对象，因此 User 类要实现 Serializable 接口，代码如下：

```java
public class User implements Serializable {
    private static final long serialVersionUID = -8418444131403945274L;
    private String userName;
    private String passWord;
    private String email;
    private int level;
    private Date regDate;
    public User(String userName, String passWord, String email,int level,Date regDate) {
        this.userName = userName;
        this.passWord = passWord;
        this.email = email;
        this.level = level;
        this.regDate = regDate;
    }
    public String getEmail() {
        return email;
    }
    public void setEmail(String email) {
        this.email = email;
    }
    public String getPassWord() {
        return passWord;
    }
```

```java
        public void setPassWord(String passWord) {
            this.passWord = passWord;
        }
        public String getUserName() {
            return userName;
        }
        public void setUserName(String userName) {
            this.userName = userName;
        }
        public int getLevel() {
            return level;
        }
        public void setLevel(int level) {
            this.level = level;
        }
        public Date getRegDate() {
            return regDate;
        }
        public void setRegDate(Date regDate) {
            this.regDate = regDate;
        }
        public String toString(){
            return userName+":"+passWord+":"+email +":" + level + ":" + regDate;
        }
    }
```

User 类中的属性分别对应 user 表中的字段，这里注册日期的类型 Date 是 java.util.Date。添加 toString()方法是为了在测试程序时，方便输出 User 对象的信息。

2. 用户管理类

在 user 包中创建用户管理类 UserDao，完成向 user 表添加用户以及在 user 表中根据用户名和密码查找用户，代码如下：

```java
    public class UserDao {
        Connection conn = null;
        Statement st = null;
        PreparedStatement pstmt;
        ResultSet rs = null;
        public UserDao(){
            conn = new DBConnection().getConn();
        }
        public boolean addUser(User user){
            String userName = user.getUserName();
            String password = user.getPassWord();
            String email = user.getEmail();
            int level = user.getLevel();
            Date regDate = user.getRegDate();
            try {
```

```java
            st = conn.createStatement();
            rs = st.executeQuery("select * from user where name='" + userName + "'");
            if(rs.next()){
                return false;
            }
            else{
                String sql="insert into user(name,password,email,level,regDate)"+
                        "values(?,?,?,?,?)";
                pstmt = conn.prepareStatement(sql);
                pstmt.setString(1,userName);
                pstmt.setString(2, password);
                pstmt.setString(3, email);
                pstmt.setInt(4, level);
                pstmt.setDate(5, new java.sql.Date(regDate.getTime()));
                pstmt.executeUpdate();
                return true;
            }
        }
        catch (SQLException e) {
            e.printStackTrace();
            return false;
        }
        finally{
            if(rs!=null){
                try {
                    rs.close();
                } catch (SQLException e) {
                    e.printStackTrace();
                }
            }
            if(st!=null){
                try {
                    st.close();
                } catch (SQLException e) {
                    e.printStackTrace();
                }
            }
        }
    }
    public User getUser(String userName, String password) {
        try {
            st = conn.createStatement();
            String sql = "select * from user where name='" + userName +
                    "' and password='" + password + "'";
            rs = st.executeQuery(sql);
            if(rs.next()){
```

```java
                    String email = rs.getString("email");
                    int level = rs.getInt("level");
                    Date regDate = rs.getDate("regDate");
                    return new User(userName,password,email,level,regDate);
                }
                else{
                    return null;
                }
            }
            catch (SQLException e) {
                e.printStackTrace();
                return null;
            }
            finally{
                if(rs!=null){
                    try {
                        rs.close();
                    } catch (SQLException e) {
                        e.printStackTrace();
                    }
                }
                if(st!=null){
                    try {
                        st.close();
                    } catch (SQLException e) {
                        e.printStackTrace();
                    }
                }
            }
        }
    }
```

addUser(User user)方法向用户表添加一个用户。因为表中不允许有同名的用户，因此在用户表中查找和 user 同名的记录，如果找到，则不能添加，返回 false；如果找不到，则将 user 对象代表的用户数据添加到用户表中，返回 true。

getUser(String userName, String password)方法在用户表中查找参数指定的用户名和密码对应的记录，如果找到，则由这条记录的数据创建 User 对象并返回；如果找不到，则返回 null。

3．测试用户类和用户管理类

完成了 User 类和 UserDao 类的构建之后，可以添加一个临时的测试类，测试 UserDao 类添加用户和查找用户的功能，例如下面的程序。

```java
public class Test {
    public static void main(String[] args) {
        UserDao dao = new UserDao();
        User user = new User("test3","aaa","tetsts",1,new Date());
        if(dao.addUser(user))
            System.out.println("添加成功！");
```

```
            else
                System.out.println("添加失败！");
            User user2=dao.getUser("test3", "aaa");
            if(user2!=null){
                System.out.println("找到该用户！");
                System.out.println(user2);
            }else{
                System.out.println("未找到该用户！");
            }

            User user3=dao.getUser("test3", "dddd");
            if(user3!=null){
                System.out.println("找到该用户！");
            }else{
                System.out.println("未找到该用户！");
            }
        }
    }
```

根据测试结果，对照数据库中的 user 表，如果有不合理的地方，则需要修改 User 类或 UserDao 类，测试成功后，可将测试类 Test 删除。

3.3 用户注册和登录

在客户端提供一个注册和登录界面，用户填好数据后向服务器发送注册或登录命令，服务器处理后，将处理结果通知客户端。运行界面如图 3.2 所示。

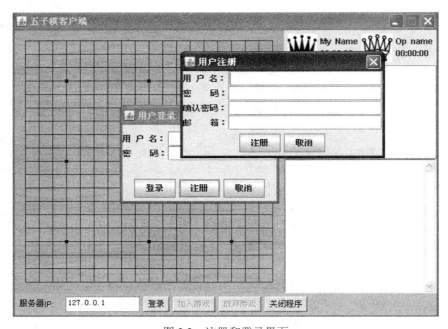

图 3.2 注册和登录界面

在客户端界面单击"登录"按钮,出现"用户登录"对话框,可以填写用户名和密码,单击"登录"按钮,向服务器发送 login 命令。如果用户还没有注册,可以单击"注册"按钮,在出现的"用户注册"对话框中输入相关信息后,单击"注册"按钮,向服务器发送 register 命令。

客户端与服务器之间发送命令的流程如图 3.3 所示。

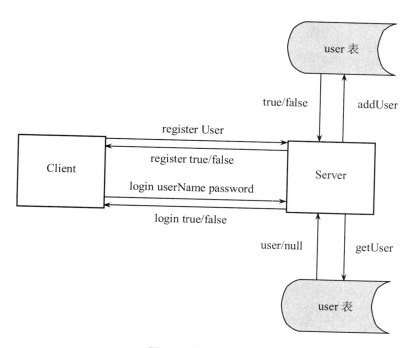

图 3.3 注册/登录流程

注册时,客户端发送 register 命令,参数是 User 对象,服务器通过 UserDao 类向用户表添加记录,如果成功,向客户端发送参数为 true 的 register 命令;如果失败,向客户端发送参数为 false 的 register 命令。

登录时,客户端发送 login 命令,参数是用户名和密码,服务器通过 UserDao 类从用户表中查找用户名和密码指定的记录,如果找到,登录成功,向客户端发送参数为 true 的 login 命令;如果没找到,登录失败,向客户端发送参数为 false 的 login 命令。

3.3.1 准备工作

1. 在 Command 类中添加命令常量

在 Command 类中添加以下两个命令常量:

```
public static final String REGISTER = "register";
public static final String LOGIN = "login";
```

2. 修改 PanelControl 类

将 PanelControl 类中的"连接主机"按钮改为"登录"按钮,即将 connectButton 改为 loginButton,将"连接主机"改为"登录",然后将程序中的所有 connectButton 替换为 loginButton。

3. FiveClient 类中添加方法

由于在登录对话框类中要用到 FiveClient 类中的 Communication 属性 c，而这两个类不在一个包中，因此需要在 FiveClient 类中添加 getC()方法，以便在其他类中获取 Communication 对象，代码如下：

```
public Communication getC(){
    return c;
}
```

3.3.2 用户登录

1. 在 Communication 类中添加登录方法

由于"连接服务器"已经被"登录"代替了，因此可以删除 FiveClient 类中的 connect()方法。将 Communication 类中的 connect()方法改为 login()方法，代码如下：

```
public boolean login(String ip, String userName, String password) {
    try {
        s = new Socket(ip,FiveServer.TCP_PORT);
        dis = new DataInputStream(s.getInputStream());
        dos = new DataOutputStream(s.getOutputStream());
        dos.writeUTF(Command.LOGIN + ":" + userName + ":" + password);
        String msg = dis.readUTF();
        String[] words = msg.split(":");
        if(words[0].equals(Command.LOGIN) && words[1].equals("true")){
            fc.isConnected = true;
            new ReceiveThread(s).start();
            fc.control.exitGameButton.setEnabled(true);
            fc.control.loginButton.setEnabled(false);
            fc.control.joinGameButton.setEnabled(true);
            fc.control.cancelGameButton.setEnabled(false);
            return true;
        }
        else{
            s.close();
            return false;
        }
    } catch (UnknownHostException e) {
        e.printStackTrace();
    } catch (IOException e) {
        if(s!=null){
            try {
                s.close();
            } catch (IOException e1) {
                e1.printStackTrace();
            }
        }
        e.printStackTrace();
    }
```

            return false;
    }
连接服务器后，向服务器发送 login 命令，然后等待服务器返回的结果，如果返回"login:true"，说明登录成功，为相关变量设置值，创建接收消息的线程并启动，设置按钮的状态；如果返回"login:false"，说明登录失败。

2. 登录对话框

在 user 包中创建登录对话框类 DialogLogin，代码如下：

```java
public class DialogLogin extends JDialog implements ActionListener{
    JTextField tfUserName = new JTextField(14);
    JTextField tfPassword = new JTextField(14);
    JButton jbLogin = new JButton("登录");
    JButton jbRegister = new JButton("注册");
    JButton jbCancel = new JButton("取消");
    String ip;
    FiveClient fc;
    public DialogLogin(Window parent, String ip) {
        super(parent, "用户登录",Dialog.ModalityType.APPLICATION_MODAL);
        fc = (FiveClient) parent;
        this.ip = ip;
        createGUI();
    }
    public void actionPerformed(ActionEvent e) {
        String str = e.getActionCommand();
        if ("登录".equals(str)) {
            String userName = tfUserName.getText();
            String password = tfPassword.getText();
            if((userName==null)||(userName.isEmpty()) ||
                    (password==null)||(password.isEmpty())){
                JOptionPane.showMessageDialog(this, "各项数据不能为空！");
                return;
            }
            else {
                if(fc.getC().login(ip, userName, password)){
                    JOptionPane.showMessageDialog(this, "登录成功！");
                    this.dispose();
                }
                else{
                    JOptionPane.showMessageDialog(this, "用户名或密码不符！");
                }
            }
        }
        else if ("注册".equals(str)) {
            //DialogRegister rd = new DialogRegister(this,ip);
        }
```

```java
            else if("取消".equals(str)){
                this.dispose();
            }
        }
        public void createGUI() {
            this.setLayout(new BorderLayout());
            JPanel jpWest = new JPanel();
            JPanel jpCenter = new JPanel();
            JPanel jpSouth = new JPanel();
            jpWest.setLayout(new GridLayout(3, 1));
            jpCenter.setLayout(new GridLayout(3, 1));
            jpSouth.setLayout(new FlowLayout());
            jpWest.add(new JLabel("用  户  名："));
            jpWest.add(new JLabel("密      码："));
            jpCenter.add(tfUserName);
            jpCenter.add(tfPassword);
            this.add(new JPanel(),BorderLayout.NORTH);
            this.add(jpWest,BorderLayout.WEST);
            this.add(jpCenter,BorderLayout.CENTER);
            jpSouth.add(jbLogin);
            jpSouth.add(jbRegister);
            jpSouth.add(jbCancel);
            jbLogin.addActionListener(this);
            jbRegister.addActionListener(this);
            jbCancel.addActionListener(this);
            this.add(jpSouth,BorderLayout.SOUTH);
            this.setDefaultCloseOperation(DISPOSE_ON_CLOSE);
            this.setLocation(450,250);
            this.pack();
            this.setVisible(true);
        }
    }
```

由于对话框本身负责事件监听，因此 DialogLogin 类实现了 ActionListener 接口。为了便于看清程序的逻辑结构，我们将创建界面的代码放在了一个单独的方法 createGUI()中。在构造方法中设置对话框的父窗口（由参数指定）、标题以及显示模式，然后调用 createGUI()方法创建对话框界面并显示。

在 actionPerformed()方法中处理三个按钮的单击事件，如果是"登录"按钮，则获取编辑框中的用户名和密码，如果用户名和密码都不为空，则调用 Communication 类中的 login()方法登录，如返回 true，则登录成功，如返回 false，则登录失败。

如果是"注册"按钮，则显示注册对话框，完成注册功能；如果是"取消"按钮，则使对话框消失。

3. 显示登录对话框

在 fiveClient 类中修改"登录"按钮的监听程序段，显示登录对话框，代码如下：

```
        else if(e.getSource() == control.loginButton){
            c = new Communication(FiveClient.this);
            DialogLogin dr = new DialogLogin(FiveClient.this, FiveClient.this.control.inputIP.getText());
}
```
在创建 DialogLogin 对象时，要通过构造方法的参数指定对话框的父窗口和服务器的 ip 地址。

4. 服务器处理 login 命令

由于现在已经不需要服务器为用户起名了，因此可以删除服务器类的静态成员 clientNameNum，然后修改 startServer()方法，修改后的代码如下：

```
public void startServer(){
    try {
        ss = new ServerSocket(TCP_PORT);
        while(true){
            Socket s = ss.accept();
            InputStream is = s.getInputStream();
            OutputStream os = s.getOutputStream();
            DataInputStream dis = new DataInputStream(is);
            DataOutputStream dos = new DataOutputStream(os);
            String msg = dis.readUTF();
            String[] words = msg.split(":");
            if(words[0].equals(Command.LOGIN)){
                String userName = words[1];
                String password = words[2];
                UserDao ud = new UserDao();
                if(ud.getUser(userName, password) != null){
                    dos.writeUTF(Command.LOGIN + ":true");
                    clientNum++;
                    Client c = new Client(userName, s);
                    clients.add(c);
                    lStatus.setText("连接数" + clientNum);
                    taMessage.append(s.getInetAddress().getHostAddress()+" "+userName+"\n");
                    tellName(c);
                    addAllUserToMe(c);
                    addMeToAllUser(c);
                    new ClientThread(c).start();
                }
                else{
                    dos.writeUTF(Command.LOGIN+ ":false");
                    s.close();
                }
            }
        }
    } catch (IOException e) {
        e.printStackTrace();
    }
}
```

现在客户端连接服务器有两个目的：注册和登录，因此服务器在接到客户端连接请求时，要分别处理这两种情况。如果是 login 命令，则将用户名和密码取出，利用 UserDao 的 getUser() 方法在 user 表中查找指定的用户，如果返回值不为空，则登录成功，向客户端发送登录成功的命令，然后进行相应的处理，创建接收该客户端命令的线程并启动。为了不改变以前的程序结构，我们仍通过 tellName() 方法向客户端发送客户名，当然这个客户名就是客户端传过来的名字。如果登录失败，则将该 Socket 关闭，向客户端发送登录失败的命令。

3.3.3 用户注册

与登录后就准备下棋不同，注册是一个相对独立的功能，因此对于注册功能，我们不使用 Communication 类与服务器通信，而是在注册对话框类中添加自己的 Socket 属性实现与服务器的连接。

1. 注册对话框类

在 user 包中创建注册对话框 DialogRegister 类，代码如下：

```java
public class DialogRegister extends JDialog implements ActionListener {
    JTextField tfUserName = new JTextField(20);
    JTextField tfPassword = new JTextField(20);
    JTextField tfRePassword = new JTextField(20);
    JTextField tfEmail = new JTextField(20);
    JButton jbRegister = new JButton("注册");
    JButton jbCancel = new JButton("取消");
    String ip;
    public DialogRegister(Window parent, String ip) {
        super(parent, "用户注册", Dialog.ModalityType.APPLICATION_MODAL);
        this.ip = ip;
        createGUI();
    }
    public void actionPerformed(ActionEvent e) {
        String str = e.getActionCommand();
        if ("注册".equals(str)) {
            String userName = tfUserName.getText();
            String password = tfPassword.getText();
            String rePassword = tfRePassword.getText();
            String email = tfEmail.getText();
            if((userName==null)||(userName.isEmpty()) || (password==null)||
                (password.isEmpty()) || (rePassword==null)||
                (rePassword.isEmpty()) || (email==null)||(email.isEmpty())){
                JOptionPane.showMessageDialog(this, "各项数据不能为空！");
                return;
            }
            if (!(password.equals(rePassword))) {
                JOptionPane.showMessageDialog(this, "两次密码不一致！");
                return;
```

```java
            }
            else {
                if(register(new User(userName,password,email,1,new Date()))){
                    JOptionPane.showMessageDialog(this, "注册成功！");
                }
                else{
                    JOptionPane.showMessageDialog(this, "注册失败！可能重名。");
                }
            }
        }
        else if("取消".equals(str)){
            this.dispose();
        }
    }
}
public boolean register(User u){
    Socket s = null;
    InputStream is = null;
    OutputStream os = null;
    DataInputStream dis = null;
    DataOutputStream dos = null;
    ObjectOutputStream oos = null;
    try {
        s = new Socket(ip, FiveServer.TCP_PORT);
        is = s.getInputStream();
        os = s.getOutputStream();
        dis = new DataInputStream(is);
        dos = new DataOutputStream(os);
        dos.writeUTF(Command.REGISTER);
        oos = new ObjectOutputStream(os);
        oos.writeObject(u);
        String msg = dis.readUTF();
        if(msg.equals(Command.REGISTER + ":true")){
            return true;
        }
        else{
            return false;
        }
    } catch (UnknownHostException e) {
        e.printStackTrace();
        return false;
    } catch (IOException e) {
        e.printStackTrace();
        return false;
```

```java
            }
            finally{
                try {
                    oos.close();
                    dos.close();
                    dis.close();
                    s.close();
                } catch (IOException e) {
                    e.printStackTrace();
                }
            }
        }
        public void createGUI() {
            this.setLayout(new BorderLayout());
            JPanel jpWest = new JPanel();
            JPanel jpCenter = new JPanel();
            JPanel jpSouth = new JPanel();
            jpWest.setLayout(new GridLayout(4, 1));
            jpCenter.setLayout(new GridLayout(4, 1));
            jpSouth.setLayout(new FlowLayout());
            jpWest.add(new JLabel("用    户    名："));
            jpWest.add(new JLabel("密             码："));
            jpWest.add(new JLabel("确认密码："));
            jpWest.add(new JLabel("邮             箱："));
            jpCenter.add(tfUserName);
            jpCenter.add(tfPassword);
            jpCenter.add(tfRePassword);
            jpCenter.add(tfEmail);
            this.add(jpWest,BorderLayout.WEST);
            this.add(jpCenter,BorderLayout.CENTER);
            jpSouth.add(jbRegister);
            jpSouth.add(jbCancel);
            jbRegister.addActionListener(this);
            jbCancel.addActionListener(this);
            this.add(jpSouth,BorderLayout.SOUTH);
            this.setDefaultCloseOperation(DISPOSE_ON_CLOSE);
            this.setLocation(450,250);
            this.pack();
            this.setVisible(true);
        }
    }
```

在 actionPerformed()方法中,如果是"注册"按钮的事件,则先检查用户名、密码、重复密码,以及邮箱是否为空,如果都非空再检查两次密码输入是否一致,如果一致,则调用 register()

方法注册，如果方法返回 true，表示注册成功，如果返回 false，表示注册失败。

在 register()方法中，首先连接服务器，随后向服务器发送 register 命令，接着再发一个 User 对象，之后再从服务器中读取字符串，如果是"register:true"，表示注册成功，如果是"register:false"，表示注册失败。

在登录对话框类中，通过单击"注册"按钮显示注册对话框。将原来类中注释掉的一行的注释去掉。

```
else if ("注册".equals(str)) {
    DialogRegister rd = new DialogRegister(this,ip);
}
```

2. 服务器处理注册命令

在服务器类的 startServer()方法中，与登录代码并列，加入注册的处理代码，如下：

```
if(words[0].equals(Command.REGISTER)){
    ObjectInputStream ois = null;
    try {
        ois = new ObjectInputStream(is);
        User u = (User) ois.readObject();
        UserDao ud = new UserDao();
        if(ud.addUser(u)){
            dos.writeUTF(Command.REGISTER + ":true");
        }
        else{
            dos.writeUTF(Command.REGISTER + ":false");
        }
    } catch (ClassNotFoundException e) {
        e.printStackTrace();
    }
    finally{
        dis.close();
        ois.close();
        dos.close();
        s.close();
    }
}
```

如果服务器收到 register 命令，则继续读一个 User 对象，然后调用 UserDao 类的 addUser()方法向数据库中添加一个记录，如果返回值为 true，表明添加成功，向客户端发送"register:true"，否则向客户端发送"register:false"。

3.4 记录棋局和棋谱

下棋结束后，将棋局信息保存到数据库中，将棋谱信息保存到棋谱文件中。这里我们只记录连成五子赢棋后的棋局，没有记录认输和超时的情况，方法都是一样的。

记录棋局和棋谱的流程如图 3.4 所示，当客户端向服务器发送 win 命令后，服务器将棋局信息保存到数据库，同时也将棋谱信息保存到棋谱文件中。

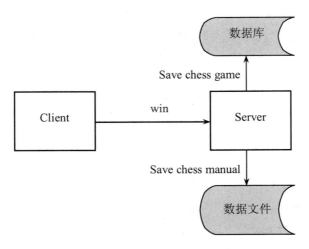

图 3.4 保存棋局和棋谱的流程

3.4.1 记录棋局

1. 棋局管理

为了方便处理棋局数据，我们创建棋局类 Game，以及保存棋局和查找棋局的类 GameDao。建立一个 game 包，在 game 包中创建 Game 类和 GameDao 类。

Game 类用于保存棋局数据，代码如下：

```java
public class Game implements Serializable {
    private static final long serialVersionUID = -1301902854300541648L;
    String bUser;           //黑方用户
    String wUser;           //白方用户
    Date date;              //下棋日期（java.util.Date）
    String winner;          //赢棋用户
    String fileName;        //保存棋谱的文件名
    public Game(String bUser, String wUser, Date date,String winner) {
        this.bUser = bUser;
        this.wUser = wUser;
        this.date = date;
        this.winner = winner;
        fileName = initFileName();
    }
    public Game(String bUser, String wUser, Date date, String winner, String fileName) {
        this.bUser = bUser;
        this.wUser = wUser;
        this.date = date;
        this.winner = winner;
        this.fileName = fileName;
    }
```

```java
private String initFileName() {
    String name;
    Calendar cal = Calendar.getInstance();
    cal.setTime(date);
    int year = cal.get(Calendar.YEAR);
    int month = cal.get(Calendar.MONTH) + 1;
    int day = cal.get(Calendar.DAY_OF_MONTH);
    int hour = cal.get(Calendar.HOUR_OF_DAY);
    int minute = cal.get(Calendar.MINUTE);
    int second = cal.get(Calendar.SECOND);
    name = bUser+ "_" +
    wUser + "_"+
    year + "_"+
    month + "_"+
    day + "_"+
    hour + "_"+
    minute + "_"+
    second + ".fiv" ;
    return   name;
}
public String getbUser() {
    return bUser;
}
public void setbUser(String bUser) {
    this.bUser = bUser;
}
public String getwUser() {
    return wUser;
}
public void setwUser(String wUser) {
    this.wUser = wUser;
}
public Date getDate() {
    return date;
}
public void setDate(Date date) {
    this.date = date;
}
public String getWinner() {
    return winner;
}
public void setWinner(String winner) {
    this.winner = winner;
}
```

```java
    public String getFileName() {
        return fileName;
    }
    public void setFileName(String fileName) {
        this.fileName = fileName;
    }
}
```

Game 类有黑方用户、白方用户、下棋日期、赢棋用户和保存棋谱的文件名等属性。当下完一局棋保存棋局时，保存棋谱文件的文件名是由下棋的两个用户名和下棋日期生成的，这时通过不包含文件名参数的构造方法创建棋局对象，在构造方法中调用 initFileName()方法获取文件名。在 initFileName()方法中根据两个对局者的用户名和棋局结束时间生成保存棋谱的文件名。

由于 Calendar 类的 get(Calendar.MONTH)方法返回值的范围是 0～11，即返回 0 表示 1 月，返回 1 表示 2 月，因此实际月份是 cal.get(Calendar.MONTH)+1。

当从数据库中读取棋局数据时，用五个参数的构造方法创建棋局对象，这时保存棋谱的文件名是由参数指定的。

由于在后面的程序中需要在网络上传递 Game 对象，因此 Game 类要实现 Serializable 接口。

GameDao 类实现向数据库的表中添加棋局和查找指定的棋局，代码如下：

```java
public class GameDao {
    Connection conn = null;
    PreparedStatement pst = null;
    Statement st = null;
    ResultSet rs;
    public GameDao(){
        conn = new DBConnection().getConn();
    }
    public boolean addGame(Game game) {
        try{
            String sql = "insert into game(gameDate,playerBlack, playerWhite, "+
                        "winner,manualFileName) values(?,?,?,?,?)";
            pst = conn.prepareStatement(sql);
            pst.setDate(1, new java.sql.Date(game.getDate().getTime()));
            pst.setString(2, game.getbUser());
            pst.setString(3, game.getwUser());
            pst.setString(4, game.getWinner());
            pst.setString(5, game.getFileName());
            pst.executeUpdate();
            return true;
        }
        catch(SQLException e){
            e.printStackTrace();
            return false;
        }
```

```java
            finally{
                if(pst!=null){
                    try{
                        pst.close();
                    }
                    catch(SQLException e){
                        e.printStackTrace();
                    }
                }
            }
        }
        public ArrayList<Game> getGame(String userName) {
            ArrayList<Game> games = new ArrayList<Game>();
            try{
                st = conn.createStatement();
                String sql = "select * from game where playerBlack= '"+userName +"'" +
                        "or playerWhite = '" +userName +"'";
                rs = st.executeQuery(sql);
                while(rs.next()){
                    String bUser = rs.getString("playerBlack");
                    String wUser = rs.getString("playerWhite");
                    java.util.Date date = new
                            java.util.Date(rs.getDate("gameDate").getTime());
                    String winner = rs.getString("winner");
                    String fileName = rs.getString("manualFileName");
                    Game g=new Game(bUser, wUser, date, winner,manualFileName);
                    games.add(g);
                }
                return games;
            }
            catch(SQLException e){
                e.printStackTrace();
                return null;
            }
            finally{
                if(rs!=null){
                    try{
                        rs.close();
                    }
                    catch(SQLException e){
                        e.printStackTrace();
                    }
                }
                if(st!=null){
                    try{
                        st.close();
```

```
                    }
                    catch(SQLException e){
                        e.printStackTrace();
                    }
                }
            }
        }
    }
```

在 GameDao 类的构造方法中,创建数据库连接。addGame()方法用于向数据库中添加一条记录,也就是一局棋的信息,成功返回 true,失败返回 false。

getGame()方法返回一个棋局链表,参数为用户名,该方法在数据库中查找参数所指定用户下过的所有棋局,将这些棋局放在链表 games 中,并将 games 作为方法的返回值。

2. 记录棋局到数据库

(1) 修改 FiveServer 的内部类 Client。

为 Client 类添加属性 chessColor、step、coordinates,分别表示该客户端执棋的颜色、下棋的手数,以及保存每一步棋子的坐标。修改后的 Client 类代码如下:

```
class Client{
    String name;
    Socket s;
    String state;          //1.ready    2.playing
    Client opponent;       //对手
    String chessColor;
    int step;
    int[][] coordinates;
    public Client(String name,Socket s) {
        this.name = name;
        this.s = s;
        this.state = "ready";
        this.opponent = null;
        step = 0;
        coordinates = new int[19*19][2];
    }
}
```

Client 对象是在客户端登录时创建的,此时还没有开始下棋,chessColor 还不能确定,在猜先后才能确定 chessColor 值。由于是 19×19 的棋盘,因此最多下 19×19 个子。我们创建一个二维数组 coordinates,每一行保存一个棋子的坐标。

(2) 猜先后给属性 chessColor 赋值。

猜先后,确定了客户端的执棋颜色,因此要修改 Command.AGREE 程序段,在猜棋部分添加代码,修改后的部分代码如下:

```
int r = (int) (Math.random()*2);   //随机分配黑棋、白棋
if(r==0){
    dos = new DataOutputStream(c.s.getOutputStream());
    dos.writeUTF(Command.GUESSCOLOR+ ":black:" + opponentName);
    dos = new DataOutputStream(c.opponent.s.getOutputStream());
```

```
            dos.writeUTF(Command.GUESSCOLOR+ ":white:" + c.name);
        c.chessColor = "black";
        c.opponent.chessColor = "white";
    }
    else
    {
        dos = new DataOutputStream(c.s.getOutputStream());
        dos.writeUTF(Command.GUESSCOLOR+ ":white:" + opponentName);
        dos = new DataOutputStream(c.opponent.s.getOutputStream());
        dos.writeUTF(Command.GUESSCOLOR+ ":black:" + c.name);
        c.chessColor = "white";
        c.opponent.chessColor = "black";
    }
```

如果自己猜到黑棋,将 chessColor 赋值为"black",对手的 chessColor 赋值为"white";如果自己猜到白棋,将 chessColor 赋值为"white",对手的 chessColor 赋值为"black"。

(3) 记录每一步棋的坐标。

每下一个棋子,服务器都会收到 GO 命令,因此可以在 GO 程序段记录下子的坐标,修改后的代码如下:

```
            else if(words[0].equals(Command.GO)){
                dos = new DataOutputStream(c.opponent.s.getOutputStream());
                dos.writeUTF(msg);
                taMessage.append(c.name + " " + msg + "\n");
                String x = words[1];
                String y = words[2];
                c.coordinates[c.step][0] = Integer.valueOf(x);
                c.coordinates[c.step][1] = Integer.valueOf(y);
                c.step++;
                c.opponent.coordinates[c.opponent.step][0] = Integer.valueOf(x);
                c.opponent.coordinates[c.opponent.step][1] = Integer.valueOf(y);
                c.opponent.step++;
            }
```

在我们的程序中,每个用户都要记录下棋过程,因此每下一步棋都要将棋子的坐标赋给自己的 coordinates 和对手的 coordinates,并将下棋手数增加 1。

(4) 记录棋局。

在 ClientThread 类中添加记录棋局的方法 recordGame(),代码如下:

```
    public void recordGame(Date date){
        String blackUser;
        String whiteUser;
        String winner;
        winner = c.name;
        if(c.chessColor.equals("black")){
            blackUser = c.name;
            whiteUser = c.opponent.name;
        }
```

```
        else{
            blackUser = c.opponent.name;
            whiteUser = c.name;
        }
        Game game = new Game(blackUser, whiteUser, date, winner);
        GameDao gd = new GameDao();
        gd.addGame(game);
    }
```

方法中首先获取黑方用户名和白方用户名，由于是赢棋后记录棋局，因此赢棋方肯定是自己，下棋时间是下棋的结束时间，通过参数获得；然后创建 Game 对象，并通过 GameDao 的 addGame()方法将 Game 对象添加到数据库中；最后在 FiveServer 类的 WIN 程序段的最后调用 recordGame()方法，将棋局信息保存到数据库中，代码如下：

```
        Date date = new Date();
        recordGame(date);
```

3.4.2 记录棋谱

1. 棋谱管理

与棋局一样，我们也要创建棋谱类 Manual，以及保存棋谱和查找棋谱的类 ManualDao。这两个类也放在 game 包中。

（1）Manual 类。

Manual 类代码如下：

```
    public class Manual extends Game implements Serializable {
        private static final long serialVersionUID = -7422915688944816730L;
        int totalStep;       //总手数
        int[][] chessList;
        public Manual(String bUser, String wUser, java.util.Date date,
                String winner, int totalStep, int[][] chessList) {
            super(bUser, wUser, date, winner);
            this.totalStep = totalStep;
            this.chessList = chessList;
        }
        public int getTotalStep() {
            return totalStep;
        }
        public void setTotalStep(int totalStep) {
            this.totalStep = totalStep;
        }
        public int[][] getChessList() {
            return chessList;
        }
        public void setChessList(int[][] chessList) {
            this.chessList = chessList;
        }
```

```java
        public static long getSerialversionuid() {
            return serialVersionUID;
        }
    }
```

棋谱信息是在棋局信息的基础上,多了一些下棋过程的信息,因此棋谱类 Manual 从棋局类 Game 继承(将棋局类作为棋谱类的父类,在面向对象的设计上不是很合理,可以采取以下几种处理方式,第一种方式是将棋谱类作为一个单独的类,包含所有需要的属性;第二种方式是在棋谱类中包含一个棋局类的属性;第三种方式是棋谱类只包含总手数和每步棋的坐标,而不包含棋局类的信息。有兴趣的读者可自行研究实现)。增加的两个属性是,下棋的总手数 totalStep 和记录棋子坐标的二维数组 chessList。

由于要向文件中写入 Manual 对象,因此 Manual 类要实现 Serializable 接口。

(2)ManualDao 类。

ManualDao 类实现棋谱写入文件和从文件中读取数据,代码如下:

```java
    public class ManualDao {
        public boolean addManual(Manual manual) {
            FileOutputStream fos=null;
            ObjectOutputStream oos=null;
            try{
                fos = new FileOutputStream("d:\\fivegame\\"+manual.getFileName());
                oos = new ObjectOutputStream(fos);
                oos.writeObject(manual);
                oos.close();
                fos.close();
                return true;
            }
            catch(IOException e){
                e.printStackTrace();
                try{
                    if(fos!=null)
                        fos.close();
                }
                catch(IOException e1){
                    e1.printStackTrace();
                }
                return false;
            }
        }
        public Manual getManual(String fileName) {
            FileInputStream fis=null;
            ObjectInputStream ois=null;
            try{
                fis = new FileInputStream("d:\\fivegame\\"+fileName);
                ois = new ObjectInputStream(fis);
```

```java
                Manual manual = (Manual) ois.readObject();
                ois.close();
                fis.close();
                return manual;
            }
            catch(FileNotFoundException e){
                e.printStackTrace();
                return null;
            }
            catch(IOException e){
                e.printStackTrace();
                return null;
            }
            catch(ClassNotFoundException e){
                e.printStackTrace();
                return null;
            }
        }
    }
```

这里我们将棋谱保存的位置固定在 d:\\fivegame 文件夹。在 addManual()方法中将一个 Manual 对象一次写入文件，在 getManual()方法中从文件中一次读取一个 Manual 对象。

2. 记录棋谱到文件

在 ClientThread 类中添加记录棋谱的方法 recordManual()，代码如下：

```java
    public void recordManual(Date date){
        String blackUser;
        String whiteUser;
        String winner;
        winner = c.name;
        if(c.chessColor.equals("black")){
            blackUser = c.name;
            whiteUser = c.opponent.name;
        }
        else{
            blackUser = c.opponent.name;
            whiteUser = c.name;
        }
        Manual manual = new Manual(blackUser, whiteUser, date, winner,c.step,c.coordinates);
        ManualDao md = new ManualDao();
        md.addManual(manual);
    }
```

recordManual()方法首先创建 Manual 对象，然后调用 ManualDao 类的 addManual()方法将棋谱写入文件。

由于是赢棋后记录棋谱，应在 FiveServer 类的 WIN 程序段的最后调用 recordManual()方法

（在 recordGame()方法之后），将棋谱保存到数据文件中，代码如下：
```
Date date = new Date();
recordGame(date);
recordManual(date);
```

3.5 查询棋局和棋谱欣赏

棋局和棋谱分别被保存到数据库和文件中之后，可以将自己下过的所有棋局查询出来，显示到一个对话框中，从中选择某个棋局，将该局棋的棋谱查找出来并进行回放。棋局和棋谱查询的流程如图 3.5 所示。

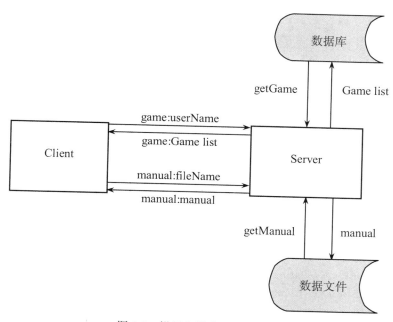

图 3.5 棋局和棋谱的查询流程

首先客户端向服务器发送 game 命令，参数为客户端的用户名，服务器收到命令后，从数据库中查找该用户下过的所有棋局，并将棋局保存在棋局链表中，然后向客户端发送 game 命令，接着继续发送查询到的棋局链表，客户端将收到的棋局显示在对话框中。

用户从对话框中选择某个棋局，向服务器发送 manual 命令，参数为保存该棋谱的文件名，服务器收到命令后，将棋谱从棋谱文件中读出来发送给客户端，客户端完成棋谱的回放。

3.5.1 查询棋局

1. 增加"棋谱欣赏"按钮

在 PanelControl 类中添加"棋谱欣赏"按钮 chessManualButton，将"棋谱欣赏"按钮排在"关闭程序"按钮之前。

在 FiveClient 类的构造方法中注册"棋谱欣赏"按钮的监听器，并将其状态设置为不可用（连接服务器之后，再将其状态设置为可用），代码如下：

```
        control.chessManualButton.addActionListener(monitor);
        control.chessManualButton.setEnabled(false);
```
在 Communication 类的 login() 方法中，如果登录成功，将 chessManualButton 按钮设置为可用状态，增加下面一行代码：
```
        fc.control.chessManualButton.setEnabled(true);
```
在 FiveClient 的内部监听器类 ActionMonitor 中，添加"棋谱欣赏"按钮的监听程序段如下：
```
        else if(e.getSource()==control.chessManualButton){
            c.getGames(myname);
        }
```
在这个程序段中调用 Communication 类中的 getGames() 方法向服务器发送 game 命令。

2. 客户端向服务器发送 game 命令

在 Command 类中添加命令常量 GAME，然后在 Communication 类中添加 getGames() 方法，代码如下：
```
        public void getGames(String myName){
            try{
                String msg = Command.GAME + ":" +myName;
                dos.writeUTF(msg);
            }
            catch(IOException e){
                e.printStackTrace();
            }
        }
```

3. 服务器接收 game 命令并处理

在服务器类的内部类 ClientThread 中添加处理 game 命令的分支，代码如下：
```
        else if(words[0].endsWith(Command.GAME)){
            ArrayList<Game> games;
            GameDao gd = new GameDao();
            String userName = words[1];
            games = gd.getGame(userName);
            DataOutputStream dos = new DataOutputStream(c.s.getOutputStream());
            dos.writeUTF(Command.GAME);
            ObjectOutputStream oos = new ObjectOutputStream(c.s.getOutputStream());
            oos.writeObject(games);
        }
```
服务器收到 game 命令后，利用 GameDao 从数据库中查询该用户下过的棋局，保存在链表 games 中，然后向客户端发送 game 命令，接着再将 games 发送给客户端。

4. 客户端接收棋局并显示

（1）显示棋局的对话框。

显示棋局的对话框如图 3.6 所示，对话框分为上下两部分，上方用 JTable 组件显示棋局，每局棋占用一行，下方是两个按钮。

图 3.6 显示棋局的对话框

对话框类 DialogGames 的代码如下：

```java
public class DialogGames extends JDialog implements ActionListener{
    private final int width = 500;
    private final int height = 400;
    private final int left=100;
    private final int top=100;
    FiveClient parent;
    JScrollPane mainPanel;
    JPanel southPanel = new JPanel();
    JTable tbGames;
    GamesTableModel model;
    JButton btnOK = new JButton("确定");
    JButton btnCancel = new JButton(" 返回");
    public DialogGames(FiveClient parent, ArrayList<Game> games)
    {
        super(parent,"选择棋局",true);
        this.parent = parent;
        setBounds(left,top ,width,height);
        model = createTableModel(games); //建立模板列表模型
        tbGames = new JTable(model);
        tbGames.setSelectionMode(ListSelectionModel.SINGLE_SELECTION);
        mainPanel = new JScrollPane(tbGames);
        southPanel.add(btnOK);
        southPanel.add(btnCancel);
        btnOK.addActionListener(this);
        btnCancel.addActionListener(this);
        add(mainPanel,BorderLayout.CENTER);
        add(southPanel,BorderLayout.SOUTH);
        setVisible(true);
    }
    public GamesTableModel createTableModel(ArrayList<Game> games){
        String[] strColumnName = new String[5];
        strColumnName[0] = "黑方";
        strColumnName[1] = "白方";
```

```java
            strColumnName[2] = "日期";
            strColumnName[3] = "胜方";
            strColumnName[4] = "文件名";
            GamesTableModel model = null;
            System.out.println(games.size());
            model = new GamesTableModel(games, strColumnName);
            return model;
        }
        public class GamesTableModel extends DefaultTableModel {
            ArrayList<Game> games;
            public GamesTableModel(ArrayList<Game> games,String[] columnNames)
            {
                super(columnNames, games.size());
                this.games = games;
            }
            public Object getValueAt(int r, int c){
                Game game = games.get(r);
                switch(c){
                case 0: return game.getbUser();
                case 1: return game.getwUser();
                case 2: return game.getDate();
                case 3: return game.getWinner();
                case 4: return game.getFileName();
                default: return null;
                }
            }
        }
        public void actionPerformed(ActionEvent e){
            if(e.getSource()==btnOK){
                int row =tbGames.getSelectedRow();
                String fileName = (String) tbGames.getValueAt(row,4);
                this.dispose();
                //通过 Communication 类向服务器发送 manual 命令
            }
            else{
                this.dispose();
            }
        }
    }
```

JTable 组件不存储自己的数据,而是从一个表格模型中获取数据。在我们的程序中,定义一个内部类 GamesTableModel 作为我们的表格模型,该类继承于 DefaultTableModel 类。在构造方法中调用父类的构造方法确定表格的列数、列标题和行数。在表格模型中,要实现 getValueAt(int r, int c)方法,以得到单元格的数据。

在 actionPerformed()方法中,实现两个按钮的监听。如果单击"返回"按钮,对话框消失;如果单击"确定"按钮,首先获取当前选择棋局的棋谱文件名,然后通过 Communication 类向服务器发送 manual 命令（这部分功能稍后实现）。

（2）FiveClient 类添加 showGames 方法。

在 FiveClient 类中添加 showGames()方法，代码如下：

```
public void showGames(ArrayList<Game> games){
    DialogGames dg = new DialogGames(this, games);
}
```

（3）Communication 类处理 game 消息。

在 Communication 类中增加处理 game 消息的分支，代码如下：

```
else if(words[0].equals(Command.GAME)){
    ArrayList<Game> games;
    ObjectInputStream ois = new ObjectInputStream(s.getInputStream());
    try{
        games = (ArrayList<Game>) ois.readObject();
        fc.showGames(games);
    }
    catch(ClassNotFoundException e){
        e.printStackTrace();
    }
}
```

客户端收到 Game 命令后继续读取 games 链表，然后调用 FiveClient 类的 showGames()方法显示棋局对话框。

3.5.2 棋谱欣赏

选择棋局对话框中的某个棋局，单击"确定"按钮，实现棋谱的读取和回放。棋谱回放对话框如图 3.7 所示。

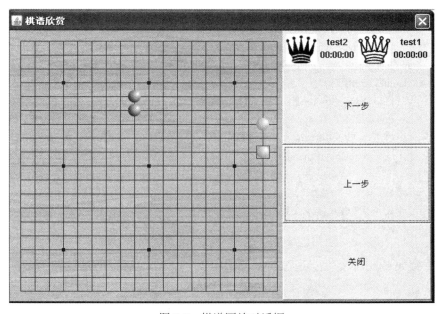

图 3.7 棋谱回放对话框

棋盘部分与主界面相同，右侧上方还是计时面板，右侧下方的三个按钮用来控制棋谱的回放以及关闭对话框。在计时面板中总是黑方在前面，我们只需要将黑方的用户名显示在前面就可以了，图标就不用设置了，同时在棋谱回放界面中也省略了计时功能（我们的程序中，在保存棋谱时，也没有保存下棋用时，因此目前无法实现显示每手棋的用时功能）。

1. 向服务器发送 manual 命令

在棋局对话框的 actionPerformed()方法中，在处理"确定"按钮事件的程序段中加入下面一行代码：

```
parent.getC().getManual(fileName);
```

其作用是调用 Communication 类的 getManual()方法，向服务器发送读取棋谱的命令。

在 Command 类中添加 MANUAL 命令常量，然后在 Communication 类中添加 getManual() 方法。

```
public void getManual(String fileName) {
    try{
        String msg = Command.MANUAL + ":" +fileName;
        dos.writeUTF(msg);
    }
    catch(IOException e){
        e.printStackTrace();
    }
}
```

manual 命令的参数是棋谱文件名。

2. 服务器接收 manual 命令并处理

在服务器的内部类 ClientThread 中添加处理 manual 命令的程序分支，代码如下：

```
else if(words[0].equals(Command.MANUAL)){
    Manual manual;
    ManualDao md = new ManualDao();
    String fileName = words[1];
    manual = md.getManual(fileName);
    if(manual!=null){
        DataOutputStream dos = new DataOutputStream(c.s.getOutputStream());
        dos.writeUTF(Command.MANUAL);
        ObjectOutputStream oos=new ObjectOutputStream(c.s.getOutputStream());
        oos.writeObject(manual);
    }
    else{
        //向客户端发送未找到文件的错误，这里省略
    }
}
```

服务器收到 manual 命令后，通过 ManualDao 类的 getManual()方法读取棋谱，如果读到棋谱，则向客户端发送 manual 命令，接着再将棋谱 manual 发送过去；如果没有读到，向客户端发送未找到棋谱文件的信息，然后客户端做相应的处理，这里暂时省略。

3. 客户端接收棋谱并显示

（1）在棋盘类中添加方法。

由于在棋谱回放对话框中需要访问棋盘类中的 chessList 属性和 chessCount 属性，为了处理方便，将棋盘类中 chessList 属性的访问权限改为 public，再添加以下两个 public 方法：

```java
public int getChessCount() {
    return chessCount;
}
public void setChessCount(int chessCount) {
    this.chessCount = chessCount;
}
```

（2）棋谱回放对话框。

在 game 包中添加棋谱回放对话框类 DialogPlayback，代码如下：

```java
public class DialogPlayback extends Dialog implements ActionListener{
    Manual manual;
    Chess[] chessList;
    PanelBoard board;
    PanelTiming timing;
    Button btNext = new Button("下一步");
    Button btPre = new Button("上一步");
    Button btCancel = new Button("关闭");
    public DialogPlayback(FiveClient parent, Manual manual){
        super(parent, "棋谱欣赏",true);
        this.manual = manual;
        chessList = new Chess[manual.getTotalStep()];
        board=new PanelBoard(null);
        timing = new PanelTiming();
        timing.setMyName(manual.getbUser());
        timing.setOpName(manual.getwUser());
        this.add(board,BorderLayout.CENTER);
        Panel east = new Panel(new BorderLayout());
        east.add(timing,BorderLayout.NORTH);
        Panel control = new Panel(new GridLayout(3,1));
        btNext.addActionListener(this);
        btPre.addActionListener(this);
        btCancel.addActionListener(this);
        control.add(btNext);
        control.add(btPre);
        control.add(btCancel);
        east.add(control,BorderLayout.CENTER);
        this.add(east,BorderLayout.EAST);
        this.setLocation(300,100);
        pack();
        this.setResizable(false);
        this.setVisible(true);
    }
```

```
            public void actionPerformed(ActionEvent e) {
                if(e.getSource()==btCancel){
                    this.dispose();
                }
                else if (e.getSource()==btNext){

                }
                else if (e.getSource()==btPre){

                }
            }
```
在 DialogPlayback 类的构造方法中完成了界面的组织,有关两个按钮的监听代码在稍后实现。

(3) 在 FiveClient 类中添加 playBack 方法。

在 FiveClient 类添加 playBack()方法,代码如下:
```
        public void playBack(Manual manual) {
            DialogPlayback dp = new DialogPlayback(this, manual);
        }
```

(4) 在 Communication 类中处理 manual 命令。

在 Communication 类中添加处理 manual 命令的分值,代码如下:
```
        else if(words[0].equals(Command.MANUAL)){
            Manual manual;
            ObjectInputStream ois = new ObjectInputStream(s.getInputStream());
            try{
                manual = (Manual) ois.readObject();
                fc.playBack(manual);
            }
            catch(ClassNotFoundException e){
                e.printStackTrace();
            }
        }
```

Communication 类收到 manual 命令后,继续将棋谱对象读取出来,调用 FiveClient 类的 playBack()方法显示棋谱回放对话框。

(5) 实现 GamePlayback 类中两个按钮的监听。

在 GamePlayback 类的 actionPerformed()方法中添加控制棋谱回放的两个按钮的监听代码如下:
```
        else if(e.getSource()==btNext){
            int step = board.getChessCount();
            if(step>=manual.getTotalStep()){
                return;
            }
            Color color=null;
            if(step%2==0){
```

```
                    color = Color.BLACK;
                }
                else{
                    color= Color.WHITE;
                }
                int col = manual.getChessList()[step][0];
                int row = manual.getChessList()[step][1];
                Chess chess = new Chess(board,col, row, color);
                board.chessList[step] = chess;
                board.setChessCount(step+1);
                board.repaint();
            }
            else if(e.getSource()==btPre){
                int step = board.getChessCount();
                step--;
                if(step<0){
                    step = 0;
                    return;
                }
                board.setChessCount(step);
                board.chessList[step] = null;
                board.repaint();
            }
```

"下一步"按钮的实现思路是,step 是当前棋盘上棋子的个数,manual.getTotalStep()是总手数,如果当前棋子数已经大于或等于总手数,则不能进行下一步。

step 从 0 开始,黑棋先行,因此 step 是偶数时就是黑棋,step 是奇数时就是白棋。然后获取棋子坐标,与下棋过程类似,创建棋子,将棋子添加到 chessList 中,重画棋盘。

"上一步"按钮的功能较简单,只要将棋盘上的棋子数减 1 就可以了,当然也要判断棋盘上还有没有棋子。

3.6 作业

1．如果将登录对话框、注册对话框在其父窗口中居中显示,应该如何处理?

2．目前的程序,用户 A 登录后,再启动一个客户端,用户 A 还可以登录,即同一个用户可以同时多次登录。如果禁止一个用户同时多次登录,应如何处理。

3．异常处理。在 ManualDao 类的 getManual()方法中,如果参数指定的棋谱文件不存在,会产生一个异常,目前程序没有很好地处理这一异常。请改进程序,处理好这个异常,能够提示用户棋谱文件不存在的信息。

4．异常处理。下棋结束后,在保存棋谱时,如果指定的文件夹"D:\fivegame"不存在,则产生异常,且不能创建棋谱文件。修改程序,如果指定的文件夹不存在,程序能够自动创建该文件夹。

实训 4 学生成绩管理系统

本实训设计一个简单的学生成绩管理系统，包括主界面、班级管理、学生管理、课程管理和成绩管理等功能。主界面如图 4.1 所示。

图 4.1 学生成绩管理系统主界面

主窗口的菜单条中有学生管理、课程管理和成绩管理三个菜单，我们将班级管理和学生管理放在了同一个菜单中，课程管理菜单中只有一项，就是课程管理，成绩管理菜单里则包括成绩录入、成绩修改、成绩查询等功能。

下面首先进行系统设计，然后分别实现成绩管理系统的各项功能。

4.1 系统设计

4.1.1 需求分析

本实训设计的学生成绩管理系统是一个简化的系统，功能包括班级信息的录入、修改、删除，学生信息的录入、修改、删除，课程信息的录入、修改、删除，成绩信息的录入、修改、查询等。选课等其他一些相关功能，本系统没有涉及。

1. 班级管理

班级管理包括班级信息的录入、修改和删除，我们将这三个功能放在同一个对话框中实现，如图 4.2 所示。

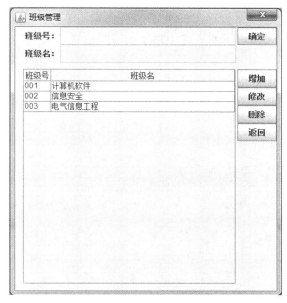

图 4.2　班级管理界面

要增加班级，在对话框中单击"增加"按钮，然后在对话框的上部输入班级号和班级名，最后单击"确定"按钮，将输入的班级添加到数据库中。对话框的下部是已有班级信息，如果要修改某个班级的信息，可以先选中该班级，然后单击"修改"按钮，则选中的班级信息出现在对话框的上部，修改后单击"确定"按钮，修改的信息就保存到数据库中。如果要删除某个班级，只要选中该班级，然后单击"删除"按钮即可。

2. 学生管理

学生管理包括学生信息的录入、修改和删除，我们也将这三个功能放在同一个对话框中实现，如图 4.3 所示。

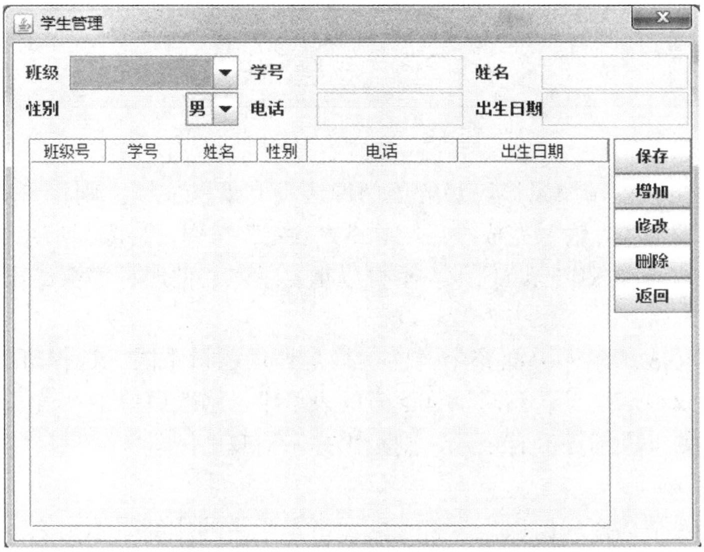

图 4.3　学生管理界面

学生管理界面的操作与班级管理界面类似,只是增加了一个班级的选择功能。对学生的任何操作都要先选择一个班级,下面表格中显示的就是所选班级的学生。

3. 课程管理

课程管理与班级管理的功能和操作一模一样,只是将班级换成了课程。

4. 成绩录入与修改

(1) 成绩录入。

成绩录入界面如图 4.4 所示。

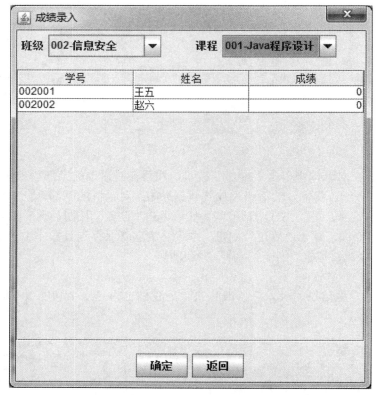

图 4.4　成绩录入对话框

首先选择班级和课程,如果该班级的该门课还没有录入成绩,则将该班级的学生名单导入到对话框下方的表格中,输入完成绩后,单击"确定"按钮,将成绩信息保存到数据库的成绩表中。如果选择的课程成绩已经录入,不可以重复录入,此时程序会给出提示信息。

(2) 成绩修改。

成绩修改界面与成绩录入界面完全一样,只是处理逻辑不同。选择班级和课程后,如果成绩已经录入则将成绩信息导入到对话框下方的表格中,修改后单击"确定"按钮,将数据库中的成绩表更新。如果成绩还没有录入,则给出提示信息。

5. 查询课程成绩

查询课程成绩就是选择一门课程,将该门课所有学生的成绩显示出来,如图 4.5 所示。

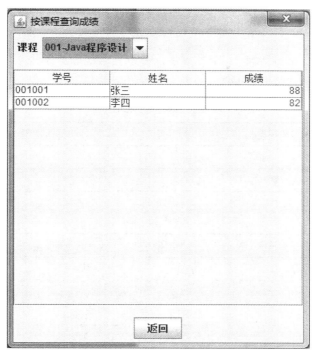

图 4.5 查询课程成绩对话框

6. 查询学生成绩

查询学生成绩是将指定学生的所有课程的成绩显示出来，如图 4.6 所示。

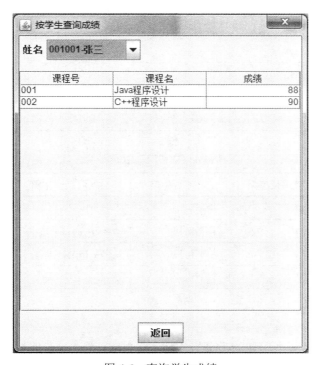

图 4.6 查询学生成绩

4.1.2 数据库设计

根据系统的功能，需要设计 4 个数据表，分别是班级表、学生表、课程表和成绩表。各表包含的字段如表 4.1 至表 4.4 所示。

表 4.1　班级表（Class）

字段名	类型（长度）	可否为空	是否主键	含义
id	CHAR(3)	不可以	是	班级号
name	VARCHAR(30)	不可以	不是	班级名

表 4.2　学生表（Student）

字段名	类型（长度）	可否为空	是否主键	含义
id	CHAR(6)	不可以	是	学号
name	CHAR(10)	不可以	不是	姓名
gender	CHAR(2)	不可以	不是	性别
tel	CHAR(11)	不可以	不是	电话
birthday	DATE	不可以	不是	出生日期
class	CHAR(3)	不可以	不是	班级号

表 4.3　课程表（Course）

字段名	类型（长度）	可否为空	是否主键	含义
id	CHAR(3)	不可以	是	课程号
name	VARCHAR(30)	不可以	不是	课程名
schoolHour	INT	不可以	不是	学时
credit	INT	不可以	不是	学分

表 4.4　成绩表（Score）

字段名	类型（长度）	可否为空	是否主键	含义
class_id	CHAR(3)	不可以	不是	班级号
stu_id	CHAR(6)	不可以	与 course_id 作为联合主键	学号
course_id	CHAR(3)	不可以	与 stu_id 作为联合主键	课程号
score	INT	不可以	不是	成绩

4.1.3 类的设计

根据类的功能，系统中主要的类分为实体类、DAO 类、对话框类。将这些类分别放在不同的包中。

1. 实体类

4 个实体类对应 4 个数据库表，分别是 ClassEntity 类、Student 类、Course 类和 Score 类。

2. 数据访问类

对应 4 个数据库表，设计 4 个数据访问类，分别是 ClassDao 类、StudentDao 类、CourseDao 类和 ScoreDao 类，实现对 4 个数据库的各种操作。

3. 对话框类

成绩管理的具体功能是通过对话框实现的，包括班级管理类 ClassManage、学生管理类 StudentManage、课程管理类 CourseManage、成绩对话框类 ScoreDialog、课程成绩查询类 ScoreQueryByCourse、学生成绩查询类 ScoreQueryByStudent。

4. 工具类

系统提供两个工具类，一个是数据库连接类 DBConnection，用于在任何时刻方便地连接数据库；另一个是 CreateDatabase 类，用于创建系统中用到的数据库和 4 个数据表。

5. 主窗口类

主窗口类 MainFrame 从 JFrame 类继承，通过菜单调用各部分功能。

4.2 工具类

两个工具类分别是数据库连接类 DBConnection 和创建数据库的类 CreateDatabase。

4.2.1 DBConnection 类

首先在 Eclipse 中创建一个名称为 Score 的 Java Project，然后在项目 Score 所在的文件夹中建立配置文件 database.properties，为数据库连接提供参数。这里我们使用的是 MySQL 数据库，配置文件内容如下：

```
driver=com.mysql.jdbc.Driver
url=jdbc:mysql://localhost:3306/Score
user=root
password=1234
```

其中后两行要根据自己的 MySQL 设置正确的用户名和密码。

为了在 Java 程序中操作数据库中的数据，需要配置数据库驱动程序 JDBC。具体方法参见 3.1.2 节中的介绍。

最后创建 util 包，在 util 包中创建 DBConnection 类，代码如下：

```java
public class DBConnection {
    //利用静态代码块，在工具类被加载时即执行配置文件的读取
    private static Properties props = new Properties();
    static {
        try {
            props.load(new FileInputStream("database.properties"));
        } catch (FileNotFoundException e) {
            e.printStackTrace();
        } catch (IOException e) {
            e.printStackTrace();
        }
```

```java
        }
        public static Connection getConnection(){
            Connection con = null;
            try {
                String driver = props.getProperty("driver");
                String url = props.getProperty("url");
                String username = props.getProperty("user");
                String password = props.getProperty("password");
                Class.forName(driver);
                con = DriverManager.getConnection(url, username, password);
            } catch (ClassNotFoundException e) {
                System.out.println("failed to register driver.");
                e.printStackTrace();
            } catch (SQLException e) {
                System.out.println("failed to execute sql.");
                e.printStackTrace();
            }
            return con;
        }
        public static void closeConnection(ResultSet rs, Statement st, Connection conn) {
            if (rs != null) {
                try {
                    rs.close();
                } catch (SQLException e) {
                    e.printStackTrace();
                }
            }
            if (st != null) {
                try {
                    st.close();
                } catch (SQLException e) {
                    e.printStackTrace();
                }
            }
            if (conn != null) {
                try {
                    conn.close();
                } catch (SQLException e) {
                    e.printStackTrace();
                }
            }
        }
        public static void main(String[] args) {
            Connection con = getConnection();
            System.out.println(con);
        }
    }
```

在静态代码段中通过 Properties 读入配置文件中的数据，方法 getConnection()返回数据库的连接，方法 closeConnection()关闭数据库连接（如果有打开的 ResultSet 和 Statement，也同时将它们关闭）。

最后一个主方法输出获得的 Connection 对象，调试成功后可将主方法删除。

4.2.2　CreateDatabase 类

CreateDatabase 类用于创建数据库和其中的数据表，方便对数据库的初始化，以及将程序拷贝到其他计算机时，快速创建数据库。

在 util 包中创建 CreateDatabase 类，代码如下：

```java
public class CreateDatabase {
    public static void main(String[] args) {
        Connection conn=null;
        Statement stmt = null;
        Properties props = new Properties();
        try {
            props.load(new FileInputStream("database.properties"));
        } catch (FileNotFoundException e) {
            e.printStackTrace();
        } catch (IOException e) {
            e.printStackTrace();
        }
        try {
            String driver = props.getProperty("driver");
            String url = "jdbc:mysql://localhost:3306/";
            String username = props.getProperty("user");
            String password = props.getProperty("password");
            Class.forName(driver);
            conn = DriverManager.getConnection(url, username, password);
            stmt = conn.createStatement();
        }
        catch (ClassNotFoundException e) {
            e.printStackTrace();
            System.exit(-1);
        }
        catch (SQLException e) {
            e.printStackTrace();
            try {
                conn.close();
            }
            catch (SQLException e1) {
                e1.printStackTrace();
                System.exit(-1);
            }
        }
        createDatabase(stmt, "Score");
```

```java
            try {
                stmt.close();
                conn.close();
            }
            catch (SQLException e) {
                e.printStackTrace();
            }
        }
        public static void createDatabase(Statement stmt, String dbName){
            try {
                stmt.executeUpdate("drop database " + dbName);
                stmt.executeUpdate("create database " + dbName);
                stmt.executeUpdate("use " + dbName);
                String sql;
                sql = "create table class(id CHAR(3) not null primary key," +
                            "name VARCHAR(30) not null)";
                stmt.executeUpdate( sql );
                sql = "create table student(id CHAR(6) not null primary key," +
                            "name CHAR(10) not null, " +
                            "gender CHAR(2) not null," +
                            "tel CHAR(11) not null," +
                            "birthday date not null," +
                            "class CHAR(3) not null)";
                stmt.executeUpdate(sql );
                sql = "create table course(id CHAR(3) not null primary key," +
                            "name    VARCHAR(30) not null, " +
                            "schoolHour int not null, " +
                            "credit int not null)";
                stmt.executeUpdate( sql );
                sql = "create table score(class_id char(3) not null, " +
                            "stu_id CHAR(6) not null," +
                            "course_id    CHAR(3) not null, " +
                            "score int not null, " +
                            "PRIMARY KEY(stu_id,course_id) )";
                stmt.executeUpdate( sql );
            }
            catch (SQLException e) {
                e.printStackTrace();
            }
        }
    }
```

在主方法中读取配置文件,并创建数据库连接和 Statement 对象,然后调用 createDatabase() 方法创建数据库和 4 个数据表。

需要注意的是,这里使用的 url 并不是从配置文件中读取的,因为配置文件中指定的数据库是 Score,此时还没有 Score 数据库,因此在 url 中还不能包含数据库的名字。

4.3 实体类

对应 4 个数据库表创建 4 个实体类，分别是班级实体类 ClassEntity、学生实体类 Student、课程实体类 Course 和成绩实体类 Score。

在项目 Score 中建立 entity 包，然后在 entity 包中创建 4 个实体类。

4.3.1 班级实体类 ClassEntity

班级实体类 ClassEntity 用于保存班级数据，一个 ClassEntity 对象代表一个班级，代码如下：

```java
public class ClassEntity {
    private String id;
    private String name;
    public ClassEntity(){
        
    }
    public ClassEntity(String id, String name) {
        super();
        this.id = id;
        this.name = name;
    }
    public String getId() {
        return id;
    }
    public void setId(String id) {
        this.id = id;
    }
    public String getName() {
        return name;
    }
    public void setName(String name) {
        this.name = name;
    }
    public String toString(){
        return id + ", " + name;
    }
}
```

班级实体类除了两个属性（班级号和班级名），就是构造方法和一组 get、set 方法，还有一个 toString()方法。

4.3.2 学生实体类 Student

学生实体类 Student 用于保存学生数据，一个 Student 对象代表一个学生，代码如下：

```java
public class Student {
    private String id;
    private String name;
```

```java
    private String gender;
    private String tel;
    private Date birthday;
    private String classNo;
    public Student(){
    }
    public Student(String id, String name, String gender, String tel,
            Date birthday, String classNo) {
        super();
        this.id = id;
        this.name = name;
        this.gender = gender;
        this.tel = tel;
        this.birthday = birthday;
        this.classNo = classNo;
    }
    public String getId() {
        return id;
    }
    public void setId(String id) {
        this.id = id;
    }
    public String getName() {
        return name;
    }
    public void setName(String name) {
        this.name = name;
    }
    public String getGender() {
        return gender;
    }
    public void setGender(String gender) {
        this.gender = gender;
    }
    public String getTel() {
        return tel;
    }
    public void setTel(String tel) {
        this.tel = tel;
    }
    public Date getBirthday() {
        return birthday;
    }
    public void setBirthday(Date birthday) {
        this.birthday = birthday;
    }
```

```java
        public String getClassNo() {
            return classNo;
        }
        public void setClassNo(String classNo) {
            this.classNo = classNo;
        }
        public String toString() {
            return id + "," + name+ "," +gender + "," +tel + "," + birthday+ "," +classNo;
        }
    }
```

学生实体类包含学号、姓名、性别、联系电话、出生日期和所在班级的班级号等属性。与班级实体类类似,学生实体类的方法也是构造方法和一组 get、set 方法,另外还有一个 toString()方法。

4.3.3 课程实体类 Course

课程实体类 Course 用于保存课程数据,一个 Course 对象代表一门课程,代码如下:

```java
    public class Course {
        private String id;
        private String name;
        private int schoolHour;
        private int credit;
        public Course(){

        }
        public Course(String id, String name, int schoolHour, int credit) {
            super();
            this.id = id;
            this.name = name;
            this.schoolHour = schoolHour;
            this.credit = credit;
        }
        public String getId() {
            return id;
        }
        public void setId(String id) {
            this.id = id;
        }
        public String getName() {
            return name;
        }
        public void setName(String name) {
            this.name = name;
        }
        public int getSchoolHour() {
            return schoolHour;
```

```java
        }
        public void setSchoolHour(int schoolHour) {
            this.schoolHour = schoolHour;
        }
        public int getCredit() {
            return credit;
        }
        public void setCredit(int credit) {
            this.credit = credit;
        }
        public String toString() {
            return id + "," + name + "," + schoolHour + "," + credit;
        }
    }
```

课程实体类包含课程号、课程名、学时和学分属性，方法与班级实体类类似。

4.3.4 成绩实体类 Score

成绩实体类 Score 用于保存成绩数据，一个 Score 对象代表一个成绩，代码如下：

```java
    public class Score {
        private String classId;
        private String studentId;
        private String courseId;
        private int score;
        public Score(){
        }
        public Score(String classId, String studentId, String courseId, int score) {
            super();
            this.classId = classId;
            this.studentId = studentId;
            this.courseId = courseId;
            this.score = score;
        }
        public String getClassId() {
            return classId;
        }
        public void setClassId(String classId) {
            this.classId = classId;
        }
        public String getStudentId() {
            return studentId;
        }
        public void setStudentId(String studentId) {
            this.studentId = studentId;
        }
        public String getCourseId() {
            return courseId;
```

```
    }
    public void setCourseId(String courseId) {
        this.courseId = courseId;
    }
    public int getScore() {
        return score;
    }
    public void setScore(int score) {
        this.score = score;
    }
    public String toString() {
        return classId + "," +studentId+ "," + courseId + "," + score;
    }
}
```

成绩实体类包含班级号、学号、课程号和成绩属性，方法与前面的实体类类似。

4.4 数据访问类

设计 4 个数据访问类：ClassDao、StudentDao、CourseDao 和 ScoreDao，分别实现班级表、学生表、课程表和成绩表的各种读写操作。

创建一个 dao 包，在 dao 包中添加下面介绍的 4 个数据访问类。

4.4.1 ClassDao 类

ClassDao 类实现对班级表的增删改查等操作，代码如下：

```
public class ClassDao {
    private Connection con;
    public ClassDao() {
    }
    public int delete(String id) {
        PreparedStatement pst = null;
        try {
            con = DBConnection.getConnection();
            String sql="delete from Class where id= ?";
            pst = con.prepareStatement(sql);
            pst.setString(1, id);
            return pst.executeUpdate();
        }catch (Exception e) {
            e.printStackTrace();
        } finally {
            DBConnection.closeConnection(null, pst, con);
        }
        return 0;
    }
    public void insert(ClassEntity cn) {
```

```java
            PreparedStatement pst = null;
            try {
                    con =DBConnection.getConnection();
                    String sql = "insert into class values(?,?)";
                    pst = con.prepareStatement(sql);
                    pst.setString(1, cn.getId());
                    pst.setString(2,cn.getName());
                    pst.executeUpdate();
            } catch (Exception e) {
                    e.printStackTrace();
            } finally {
                    DBConnection.closeConnection(null, pst, con);
            }
    }
    public List<ClassEntity> selectAll() {
            Statement st=null;
            ResultSet rs = null;
            List<ClassEntity> list= new ArrayList<ClassEntity>();
            try {
                    con = DBConnection.getConnection();
                    String sql="select * from Class";
                    st = con.createStatement();
                    rs=st.executeQuery(sql);
                    while(rs.next()){
                            ClassEntity cn = new ClassEntity();
                            cn.setId(rs.getString("id"));
                            cn.setName(rs.getString("name"));
                            list.add(cn);
                    }
            }catch (SQLException e) {
                    e.printStackTrace();
            }finally{
                    DBConnection.closeConnection(rs, st, con);
            }
            return list;
    }
    public ClassEntity selectById(String id) {
            PreparedStatement pst=null;
            ResultSet rs = null;
            ClassEntity cn = null;
            try{
                    con = DBConnection.getConnection();
                    String sql = "select * from class where id =?";
                    pst = con.prepareStatement(sql);
                    pst.setString(1, id);
                    rs=pst.executeQuery();
```

```java
                    if(rs.next()){
                        cn = new ClassEntity();
                        cn.setId(rs.getString("id"));
                        cn.setName(rs.getString("name"));
                    }
                }catch (SQLException e) {
                    e.printStackTrace();
                }finally{
                    DBConnection.closeConnection(rs, pst, con);
                }
                return cn;
            }
            public void update(ClassEntity cn) {
                PreparedStatement pst = null;
                try {
                    con = DBConnection.getConnection();
                    String sql="update Class set name=?  where id= ?";
                    pst = con.prepareStatement(sql);
                    pst.setString(1, cn.getName());
                    pst.setString(2,cn.getId());
                    pst.executeUpdate();
                } catch (Exception e) {
                    e.printStackTrace();
                } finally {
                    DBConnection.closeConnection(null, pst, con);
                }
            }
            public static void main(String[] args){
                ClassDao dao = new ClassDao();
                dao.insert(new ClassEntity("002","AAAA"));
                ClassEntity cn = dao.selectById("002");
                System.out.println(cn);
                dao.insert(new ClassEntity("003","BBBB"));
                List<ClassEntity> list = dao.selectAll();
                System.out.println(list);
                dao.delete("002");
                list = dao.selectAll();
                System.out.println(list);
            }
        }
```

  类中应该提供哪些方法是根据程序的功能决定的。根据班级管理功能的需要，ClassDao 类有以下方法：方法 delete() 从班级表中删除由参数指定班级号的班级，方法 insert() 向班级表中添加由参数指定的班级，方法 selectAll() 返回班级表中所有班级列表，方法 selectById() 返回由参数指定班级号的班级，方法 update() 更新参数指定的班级（班级号是不可以修改的，只能修改班级名）。

最后添加一个主方法，用来测试 ClassDao 类的各项功能，这里提供的主方法只是一个示例，应该将所有方法都测试到，出现问题及时解决，如果等到后面用到这些类时才发现问题，就比较麻烦了。测试好了之后可以将这个主方法删除。同样后面将要介绍的另外 3 个数据访问类也应该添加一个主方法进行测试，但在本书中不再给出对后面几个类的测试，读者可以自行编写测试代码。

4.4.2 StudentDao 类

StudentDao 类实现对学生表的增、删、改、查等操作，代码如下：

```java
public class StudentDao {
    private Connection con;
    public StudentDao() {
    }
    public int delete(String id) {
        PreparedStatement pst = null;
        try {
            con = DBConnection.getConnection();
            String sql="delete from Student where id= ?";
            pst = con.prepareStatement(sql);
            pst.setString(1, id);
            return pst.executeUpdate();
        }catch (Exception e) {
            e.printStackTrace();
        } finally {
            DBConnection.closeConnection(null, pst, con);
        }
        return 0;
    }
    public void insert(Student stu) {
        PreparedStatement pst = null;
        try {
            con =DBConnection.getConnection();
            String sql = "insert into Student values(?,?,?,?,?,?)";
            pst = con.prepareStatement(sql);
            pst.setString(1, stu.getId());
            pst.setString(2,stu.getName());
            pst.setString(3,stu.getGender());
            pst.setString(4,stu.getTel());
            //java.util.Date 转换成 java.sql.Date
            pst.setDate(5, new java.sql.Date(stu.getBirthday().getTime()));
            pst.setString(6,stu.getClassNo());
            pst.executeUpdate();
        } catch (Exception e) {
            e.printStackTrace();
        } finally {
            DBConnection.closeConnection(null, pst, con);
```

```java
        }
}
public List<Student> selectAll() {
        Statement st=null;
        ResultSet rs = null;
        List<Student> list= new ArrayList<Student>();
        try {
                con = DBConnection.getConnection();
                String sql="select * from student";
                st = con.createStatement();
                rs=st.executeQuery(sql);
                while(rs.next()){
                        Student stu = new Student();
                        stu.setId(rs.getString("id"));
                        stu.setName(rs.getString("name"));
                        stu.setGender(rs.getString("gender"));
                        stu.setTel(rs.getString("tel"));
                        stu.setBirthday(rs.getDate("birthday"));
                        stu.setClassNo(rs.getString("class"));
                        list.add(stu);
                }
        }catch (SQLException e) {
                e.printStackTrace();
        }finally{
                DBConnection.closeConnection(rs, st, con);
        }
        return list;
}
public Student selectById(String id) {
        PreparedStatement pst=null;
        ResultSet rs = null;
        Student stu = null;
        try{
                con = DBConnection.getConnection();
                String sql = "select * from student where id =?";
                pst = con.prepareStatement(sql);
                pst.setString(1, id);
                rs=pst.executeQuery();
                if(rs.next()){
                        stu = new Student();
                        stu.setId(rs.getString("id"));
                        stu.setName(rs.getString("name"));
                        stu.setGender(rs.getString("gender"));
                        stu.setTel(rs.getString("tel"));
                        stu.setBirthday(rs.getDate("birthday"));
                        stu.setClassNo(rs.getString("class"));
```

```java
            }
        }catch (SQLException e) {
            e.printStackTrace();
        }finally{
            DBConnection.closeConnection(rs, pst, con);
        }
        return stu;
    }
    public List<Student> selectByClassId(String classNo) {
        PreparedStatement pst=null;
        ResultSet rs = null;
        Student stu = null;
        List<Student> list = new ArrayList<Student>();
        try{
            con = DBConnection.getConnection();
            String sql = "select * from student where class =?";
            pst = con.prepareStatement(sql);
            pst.setString(1, classNo);
            rs=pst.executeQuery();
            while(rs.next()){
                stu = new Student();
                stu.setId(rs.getString("id"));
                stu.setName(rs.getString("name"));
                stu.setGender(rs.getString("gender"));
                stu.setTel(rs.getString("tel"));
                stu.setBirthday(rs.getDate("birthday"));
                stu.setClassNo(rs.getString("class"));
                list.add(stu);
            }
        }catch (SQLException e) {
            e.printStackTrace();
        }finally{
            DBConnection.closeConnection(rs, pst, con);
        }
        return list;
    }
    public void update(Student stu) {
        PreparedStatement pst = null;
        try {
            con = DBConnection.getConnection();
            String sql="update student set name=?,  gender=?,tel=?, "
+"birthday=?, class=? where id= ?";
            pst = con.prepareStatement(sql);
            pst.setString(1, stu.getName());
            pst.setString(2,stu.getGender());
            pst.setString(3,stu.getTel());
```

```
                    //java.util.Date 转换成 java.sql.Date
                    pst.setDate(4, new java.sql.Date(stu.getBirthday().getTime()));
                    pst.setString(5,stu.getClassNo());
                    pst.setString(6,stu.getId());
                    pst.executeUpdate();
            } catch (Exception e) {
                    e.printStackTrace();
            } finally {
                    DBConnection.closeConnection(null, pst, con);
            }
        }
    }
```

与 ClassDao 类似，StudentDao 也有 delete()、insert()、selectAll()、selectById()、update() 等方法，分别完成学生表的删除记录、添加记录、返回所有学生、查询指定学号的学生，以及更新指定学生的信息（学号是不可以修改的，只能修改其他信息）等功能，另外还有一个 selectByClassId()方法，用于查询指定班级的所有学生。

4.4.3 CourseDao 类

CourseDao 类实现对课程表的增、删、改、查等操作，代码如下：

```
    public class CourseDao {
            private Connection con;
            public CourseDao() {
            }
            public int delete(String id) {
                    PreparedStatement pst = null;
                    try {
                            con = DBConnection.getConnection();
                            String sql="delete from course where id= ?";
                            pst = con.prepareStatement(sql);
                            pst.setString(1, id);
                            return pst.executeUpdate();
                    }catch (Exception e) {
                            e.printStackTrace();
                    } finally {
                            DBConnection.closeConnection(null, pst, con);
                    }
                    return 0;
            }
            public void insert(Course course) {
                    PreparedStatement pst = null;
                    try {
                            con =DBConnection.getConnection();
                            String sql = "insert into course values(?,?,?,?)";
                            pst = con.prepareStatement(sql);
                            pst.setString(1, course.getId());
```

```java
                pst.setString(2,course.getName());
                pst.setInt(3,course.getSchoolHour());
                pst.setInt(4,course.getCredit());
                pst.executeUpdate();
        } catch (Exception e) {
                e.printStackTrace();
        } finally {
                DBConnection.closeConnection(null, pst, con);
        }
    }
    public List<Course> selectAll() {
        Statement st=null;
        ResultSet rs = null;
        List<Course> list= new ArrayList<Course>();
        try {
                con = DBConnection.getConnection();
                String sql="select * from Course";
                st = con.createStatement();
                rs=st.executeQuery(sql);
                while(rs.next()){
                        Course course = new Course();
                        course.setId(rs.getString("id"));
                        course.setName(rs.getString("name"));
                        course.setSchoolHour(rs.getInt("schoolHour"));
                        course.setCredit(rs.getInt("credit"));
                        list.add(course);
                }
        }catch (SQLException e) {
                e.printStackTrace();
        }finally{
                DBConnection.closeConnection(rs, st, con);
        }
        return list;
    }
    public Course selectById(String id) {
        PreparedStatement pst=null;
        ResultSet rs = null;
        Course course = null;
        try{
                con = DBConnection.getConnection();
                String sql = "select * from course where id =?";
                pst = con.prepareStatement(sql);
                pst.setString(1, id);
                rs=pst.executeQuery();
                if(rs.next()){
                        course = new Course();
```

```java
                    course.setId(rs.getString("id"));
                    course.setName(rs.getString("name"));
                    course.setSchoolHour(rs.getInt("schoolHour"));
                    course.setCredit(rs.getInt("credit"));
                }
            }catch (SQLException e) {
                e.printStackTrace();
            }finally{
                DBConnection.closeConnection(rs, pst, con);
            }
            return course;
        }
        public void update(Course course) {
            PreparedStatement pst = null;
            try {
                con = DBConnection.getConnection();
                String sql="update course set name=?,schoolHour=?,credit=?    where id= ?";
                pst = con.prepareStatement(sql);
                pst.setString(1, course.getName());
                pst.setInt(2, course.getSchoolHour());
                pst.setInt(3, course.getCredit());
                pst.setString(4,course.getId());
                pst.executeUpdate();
            } catch (Exception e) {
                e.printStackTrace();
            } finally {
                DBConnection.closeConnection(null, pst, con);
            }
        }
    }
```

CourseDao 类与 ClassDao 类非常类似，其方法不再解释，可参考 ClassDao 类的解释。

4.4.4 ScoreDao 类

ScoreDao 类实现对成绩表的增、删、改、查等操作，代码如下：

```java
    public class ScoreDao {
        private Connection con;
        public ScoreDao() {
        }
        public void insert(Score score) {
            PreparedStatement pst = null;
            try {
                con =DBConnection.getConnection();
                String sql = "insert into score values(?,?,?,?)";
                pst = con.prepareStatement(sql);
                pst.setString(1, score.getClassId());
                pst.setString(2, score.getStudentId());
```

```java
                pst.setString(3,score.getCourseId());
                pst.setInt(4,score.getScore());
                pst.executeUpdate();
        } catch (Exception e) {
                e.printStackTrace();
        } finally {
                DBConnection.closeConnection(null, pst, con);
        }
    }
    public List<Score> selectByStudentId(String studentId) {
        Statement st=null;
        ResultSet rs = null;
        List<Score> list= new ArrayList<Score>();
        try {
                con = DBConnection.getConnection();
                //获取查询条件
                String sql="select * from score where stu_id = '" + studentId + "'";
                st = con.createStatement();
                rs=st.executeQuery(sql);
                while(rs.next()){
                        Score score = new Score();
                        score.setStudentId(rs.getString("stu_id"));
                        score.setCourseId(rs.getString("course_id"));
                        score.setScore(rs.getInt("score"));
                        list.add(score);
                }
        }catch (SQLException e) {
                e.printStackTrace();
        }finally{
                DBConnection.closeConnection(rs, st, con);
        }
        return list;
    }
    public List<Score> selectByCourseId(String courseId) {
        Statement st=null;
        ResultSet rs = null;
        List<Score> list= new ArrayList<Score>();
        try {
                con = DBConnection.getConnection();
                //获取查询条件
                String sql="select * from score where course_id = '" + courseId + "'";
                st = con.createStatement();
                rs=st.executeQuery(sql);
                while(rs.next()){
                        Score score = new Score();
                        score.setStudentId(rs.getString("stu_id"));
```

```java
                    score.setCourseId(rs.getString("course_id"));
                    score.setScore(rs.getInt("score"));
                    list.add(score);
                }
            }catch (SQLException e) {
                e.printStackTrace();
            }finally{
                DBConnection.closeConnection(rs, st, con);
            }
            return list;
    }
    public void update(Score score) {
            PreparedStatement pst = null;
            try {
                    con = DBConnection.getConnection();
                    String sql="update score set score=?   where stu_id= ? and course_id=?";
                    pst = con.prepareStatement(sql);
                    pst.setInt(1, score.getScore());
                    pst.setString(2,score.getStudentId());
                    pst.setString(3,score.getCourseId());
                    pst.executeUpdate();
            } catch (Exception e) {
                    e.printStackTrace();
            } finally {
                    DBConnection.closeConnection(null, pst, con);
            }
    }
    public List<Score> selectByClassIdCourseId(String classNo, String courseNo) {
            Statement st=null;
            ResultSet rs = null;
            List<Score> list= new ArrayList<Score>();
            try {
                    con = DBConnection.getConnection();
                    String sql="select * from score where course_id = '" + courseNo
                                                     + "' and class_id= '" + classNo + "'";
                    st = con.createStatement();
                    rs=st.executeQuery(sql);
                    while(rs.next()){
                            Score score = new Score();
                            score.setClassId(rs.getString("class_id"));
                            score.setStudentId(rs.getString("stu_id"));
                            score.setCourseId(rs.getString("course_id"));
                            score.setScore(rs.getInt("score"));
                            list.add(score);
                    }
            }catch (SQLException e) {
```

```
                e.printStackTrace();
            }finally{
                DBConnection.closeConnection(rs, st, con);
            }
            return list;
        }
    }
```

ScoreDao 类的方法如下：insert()添加一个成绩记录，update()更新一个成绩，selectByStudentId()查询某个学生的所有成绩，selectByCourseId()查询某门课程的成绩，selectByClassIdCourseId()查询某个班级某门课程的成绩。

4.5 主窗口类

成绩管理系统的主窗口主要包含一组菜单，如图 4.1 所示，通过选择菜单可以调用成绩管理系统的各种功能。

主窗口类 MainFrame 从 JFrame 类继承，直接在默认包中创建，代码如下：

```java
public class MainFrame extends JFrame {
    JMenuBar menuBar;
    JMenu[] menu;
    JMenuItem[][] menuItems;
    String[] menuNames={"学生管理","课程管理","成绩管理"};
    String[][] itemNames={{"班级管理","学生管理"},{"课程管理"},
                    {"成绩录入","成绩修改","查询课程成绩","查询学生成绩"}};
    public MainFrame(){
        menuBar = new JMenuBar();
        menu = new JMenu[menuNames.length];
        for(int i=0; i<menu.length; i++)
            menu[i] = new JMenu(menuNames[i]);
        menuItems = new JMenuItem[menu.length][];
        Monitor m = new Monitor();
        for(int i=0; i<menu.length; i++){
            menuItems[i] = new JMenuItem[itemNames[i].length];
            for(int j=0; j<menuItems[i].length; j++){
                menuItems[i][j] = new JMenuItem(itemNames[i][j]);
                menuItems[i][j].addActionListener(m);
                menu[i].add(menuItems[i][j]);
            }
        }
        for(int i=0; i< menu.length; i++){
            menuBar.add(menu[i]);
        }
        this.setJMenuBar(menuBar);
        this.setSize(600, 400);
```

```java
            this.setLocationRelativeTo(null);
            this.setDefaultCloseOperation(EXIT_ON_CLOSE);
            this.setVisible(true);
            this.setResizable(false);
        }
        class Monitor implements ActionListener{
            public void actionPerformed(ActionEvent e) {
                JMenuItem item = (JMenuItem) e.getSource();
                if(item == menuItems[0][0]){
                    new ClassManage(MainFrame.this);
                }
                if(item == menuItems[0][1]){
                    new StudentManage(MainFrame.this);
                }
                else if(item == menuItems[1][0]){
                    new CourseManage(MainFrame.this);
                }
                else if(item == menuItems[2][0]){
                    new ScoreDialog(MainFrame.this,"成绩录入", true, 1);
                }
                else if(item == menuItems[2][1]){
                    new ScoreDialog(MainFrame.this,"成绩录入", true, 2);
                }
                else if(item == menuItems[2][2]){
                    new ScoreQueryByCourse(MainFrame.this);
                }
                else if(item == menuItems[2][3]){
                    new ScoreQueryByStudent(MainFrame.this);
                }
            }
        }
    }
```

内部类 Monitor 是菜单监听器，根据不同的菜单项调用不同的功能模块，由于目前各项功能还没有实现，因此可以先将调用功能模块的语句注释掉，等后面完成一个功能模块就添加一个调用。

在构造方法中创建菜单条、菜单和菜单项，注册监听器，并将它们组织起来，添加到主窗口中。

最后在默认包中创建主程序，在 main()方法中创建主窗口对象，代码如下：

```java
public class ScoreSystem {
    public static void main(String[] args) {
        new MainFrame();
    }
}
```

4.6 班级管理

班级管理与课程管理的功能和实现方法是完全一样的,这里只给出班级管理的实现,对于课程管理部分,读者可以参照班级管理自行设计。

班级管理功能包括班级的增、删、改三项功能,界面如图 4.2 所示。界面上方两个文本框用于添加班级和修改班级,下方表格用于显示已经存在的班级。

初始时上面的两个文本框是不可编辑的,单击"增加"按钮,两个文本框变为可编辑状态,输入班级号和班级名,单击"确定"按钮,则输入的班级添加到数据表中。选中表格中的某个班级,单击"修改"按钮,则所选班级的班级号和班级名出现在上方的文本框中,修改班级名后(班级号不可修改),单击"确定"按钮,修改任务完成。要删除某个班级,只要在表格中选中该班级,再单击"删除"按钮。

创建 dialog 包,实现各种功能的对话框类都放在 dialog 包中。

表格使用 JTable 组件实现,该组件的数据来源于 TableModel,可以直接使用 JDK 提供的 DefaultTableModel。由于在班级管理、学生管理和课程管理的界面中,表格中的数据是不可以编辑的,因此我们创建一个 MyTableModel 类(创建子包 dialog.table,在这个子包中创建 MyTableModel 类),代码如下:

```java
public class MyTableModel extends DefaultTableModel {
    public MyTableModel(Object[][] tableDate,String[] columnNames)
    {
        super(tableDate, columnNames);
    }
    public boolean isCellEditable(int row,int column){
        return false;
    }
}
```

构造方法的第一个参数是一个二维数组,用于保存表格中的初始数据,第二个参数是字符串数组,用于指定表格的列标题。

重写 isCellEditable()方法,该方法返回 false,则单元格不能编辑;该方法返回 true,则单元格可以编辑。

在 dialog 包中创建 ClassManage 类,该类继承于 JDialog 类并实现 ActionListener 接口,用于监听对话框中的按钮,代码如下:

```java
public class ClassManage extends JDialog implements ActionListener {
    private JFrame parent;
    private JLabel lbClassNo;
    private JLabel lbClassName;
    private JTextField tfClassNo;
    private JTextField tfClassName;
    private JButton btOK;
    private JTable tbClass;
    private MyTableModel model;
    private JButton btAppend;
```

```java
        private JButton btEdit;
        private JButton btDelete;
        private JButton btCancel;
        private int state;      // 0. 初始状态，   1. 添加状态，   2. 修改状态
        public ClassManage(JFrame parent){
            super(parent,"班级管理",true);
            this.parent = parent;
            state = 0;
            createGUI();
        }
        private void createGUI(){
            //创建对话框上半部分
            JPanel northWestPanel = new JPanel(new GridLayout(2,1));        //左
            lbClassNo = new JLabel("        班级号：");
            lbClassName = new JLabel("        班级名：");
            northWestPanel.add(lbClassNo);
            northWestPanel.add(lbClassName);
            JPanel northCenterPanel = new JPanel(new GridLayout(2,1));      //中
            tfClassNo = new JTextField(10);
            tfClassName = new JTextField(10);
            northCenterPanel.add(tfClassNo);
            northCenterPanel.add(tfClassName);
            JPanel northEastPanel = new JPanel(new GridLayout(2,1));        //右
            btOK = new JButton("确定");
            northEastPanel.add(btOK);
            JPanel north = new JPanel(new BorderLayout(2,2));
            north.add(northWestPanel,BorderLayout.WEST);
            north.add(northCenterPanel,BorderLayout.CENTER);
            north.add(northEastPanel,BorderLayout.EAST);
            Container c = this.getContentPane();
            c.setLayout(new BorderLayout(10,10));
            c.add(north,BorderLayout.NORTH);
            //创建对话框下半部分
            tbClass = new JTable();
            initTable();
            JScrollPane listPanel = new JScrollPane(tbClass);
            btAppend = new JButton("增加");
            btEdit = new JButton("修改");
            btDelete = new JButton("删除");
            btCancel = new JButton("返回");
            JPanel buttonPanel = new JPanel(new GridLayout(4,1));
            JPanel downEastPanel = new JPanel(new BorderLayout());          //右
            buttonPanel.add(btAppend);
            buttonPanel.add(btEdit);
            buttonPanel.add(btDelete);
            buttonPanel.add(btCancel);
```

```java
                downEastPanel.add(buttonPanel,BorderLayout.NORTH);
                JPanel downPanel = new JPanel(new BorderLayout(2,0));
                downPanel.add(new JPanel(),BorderLayout.WEST);
                downPanel.add(listPanel,BorderLayout.CENTER);
                downPanel.add(downEastPanel,BorderLayout.EAST);
                downPanel.add(new JPanel(),BorderLayout.SOUTH);
                c.add(downPanel,BorderLayout.CENTER);
                tfClassNo.setEditable(false);
                tfClassName.setEditable(false);
                btOK.addActionListener(this);
                btAppend.addActionListener(this);
                btEdit.addActionListener(this);
                btDelete.addActionListener(this);
                btCancel.addActionListener(this);
                this.pack();
                this.setLocationRelativeTo(parent);
                this.setVisible(true);
            }
            private void initTable(){
                ClassDao dao = new ClassDao();
                List<ClassEntity>list = dao.selectAll();
                model = createTableModel(list); //建立模板列表模型
                tbClass.setModel(model);
                tbClass.getColumn("班级号").setPreferredWidth(40);
                tbClass.getColumn("班级名").setPreferredWidth(280);
                tbClass.setSelectionMode(ListSelectionModel.SINGLE_SELECTION);
            }
            public MyTableModel createTableModel(List<ClassEntity> list){
                String[] strColumnName = new String[2];
                strColumnName[0] = "班级号";
                strColumnName[1] = "班级名";
                Object[][] tableData = new Object[list.size()][2];
                for(int i =0; i<tableData.length; i++){
                    ClassEntity ce = list.get(i);
                    tableData[i][0] = ce.getId();;
                    tableData[i][1] = ce.getName();
                }
                return new MyTableModel(tableData, strColumnName);
            }
            public void actionPerformed(ActionEvent e){
                JButton buttone = (JButton) e.getSource();
                if(buttone==btAppend){
                    if(state==1) //如果已经是添加状态，则不处理
                        return;
                    else if(state == 2){ //如果是编辑状态，已经编辑的内容是否作废
                        if(JOptionPane.showConfirmDialog(this,
```

```java
                    "放弃当前的修改吗？")!=JOptionPane.YES_OPTION){
                return;
            }
        }
        tfClassNo.setEditable(true);
        tfClassName.setEditable(true);
        tfClassNo.setText("");
        tfClassName.setText("");
        tfClassNo.requestFocus();
        state = 1;
    }
    else if(buttone==btEdit){
        if(state==1){
            if(JOptionPane.showConfirmDialog(this,
                    "放弃当前的添加吗？")!=JOptionPane.YES_OPTION){
                return;
            }
        }
        else if(state==2){
            if(JOptionPane.showConfirmDialog(this,
                    "放弃当前的修改吗？")!=JOptionPane.YES_OPTION){
                return;
            }
        }
        int row =tbClass.getSelectedRow();
        if(row==-1){
            JOptionPane.showMessageDialog(this, "请选中一个班级！");
            return;
        }
        String classNo = (String) tbClass.getValueAt(row,0);
        String classNname = (String) tbClass.getValueAt(row,1);
        tfClassNo.setEditable(false);
        tfClassName.setEditable(true);
        tfClassNo.setText(classNo);
        tfClassName.setText(classNname);
        tfClassName.requestFocus();
        state = 2;
    }
    else if(buttone==btDelete){
        int row =tbClass.getSelectedRow();
        if(row==-1){ //选中一个班级才能删除
            JOptionPane.showMessageDialog(this, "请选中一个班级！");
            return;
        }
        String classNo = (String) tbClass.getValueAt(row,0);
        StudentDao sd = new StudentDao();
```

```java
                List<Student> list = sd.selectByClassId(classNo);
                if(list.size()!=0){ //有学生的班级不能删除
                    JOptionPane.showMessageDialog(this,
                                "该班级已经有学生，不能删除该班级！");
                    return;
                }
                if(JOptionPane.showConfirmDialog(this, "确认删除该班级？")
                                !=JOptionPane.YES_OPTION){
                    return;
                }
                if(classNo.equals(tfClassNo.getText().trim())){
                    if(state == 2){    //删除的恰好是正在修改的班级，将编辑框清空
                        tfClassNo.setText("");
                        tfClassName.setText("");
                        tfClassNo.setEditable(false);
                        tfClassName.setEditable(false);
                        state=0;
                    }
                }
                ClassDao dao = new ClassDao();
                dao.delete(classNo);
                initTable();
            }
            else if(buttone==btOK){
                if(state==0)
                    return;
                ClassDao dao = new ClassDao();
                String classNo = tfClassNo.getText().trim();
                String className = tfClassName.getText().trim();
                if(classNo.isEmpty() || className.isEmpty()){
                    JOptionPane.showMessageDialog(this,
                                "班级号和班级名不能为空，请重新输入！");
                    return;
                }
                if(state==1){
                    ClassEntity ce = dao.selectById(tfClassNo.getText().trim());
                    if(ce!=null){
                        JOptionPane.showMessageDialog(this,"班级号重复，请重新输入！");
                        return;
                    }
                    else{
                        dao.insert(new ClassEntity(classNo, className));
                        initTable();
                    }
                }
                else if(state==2){
```

```
                    dao.update(new ClassEntity(classNo, className));
                    initTable();
                }
                state = 0;
                tfClassNo.setText("");
                tfClassName.setText("");
                tfClassNo.setEditable(false);
                tfClassName.setEditable(false);
            }
            else if(buttone==btCancel){
                this.dispose();
            }
        }
    }
}
```

方法 createGUI()创建对话框中的各个组件，以及注册监听器。方法 initTable()初始化表格，使用 ClassDao 对象读取班级表，将班级信息存放在 list 中，然后调用 createTableModel()方法创建表格模型，通过 JTable 的 setModel()方法将创建好的表格模型设置给 JTable 对象 tbClass，最后设置表格每一列的宽度，并设置表格只能选中一行。

方法 createTableModel()的参数就是班级链表，在方法中首先设计好表格每列的标题 strColumnName，然后用班级链表给表格数据 tableData 赋值，最后创建 MyTableModel 对象并将该对象作为返回值。

在 actionPerformed()方法中完成各个按钮的处理，如果单击"增加"按钮，首先判断当前的状态，如果已经是添加状态，则忽略本次单击事件；如果当前是修改状态，则提示"放弃当前的修改吗？"，如果不放弃，则忽略本次单击事件。最后将编辑框设置为可编辑状态，并清空内容，将 state 设置为 1，表示进入增加状态。

单击"修改"按钮时，如果当前是增加或修改状态，也要提示"放弃当前的添加或修改吗？"，如果不放弃，则忽略本次单击事件。接下来判断是否选择了表格中的一行，如果没有选中一行，则给出提示。最后将文本框设置为可编辑状态，并将选择行的数据添加到文本框中，将 state 设置为 2，表示进入修改状态。

如果单击的是"删除"按钮，首先判断是否选择了表格中的一行，如果没有选中一行，则给出提示。然后在学生表中查找有没有该班级的学生，如果班级中已经有学生了，则不允许删除。如果确认删除该班级，还要检查一下删除的班级是不是正在修改的班级，如果是的话将文本框清空，并设置为不可编辑状态，将 state 设置为 0，表示已经离开增加或修改状态。最后从班级表中删除该班级，调用 initTable()方法重新初始化表格（保证表格中的数据与数据库中的数据一致）。

如果单击的是"确定"按钮，state 为 0，则忽略本次单击事件；否则检查班级号和班级名是否为空，如有空值则给出提示。如果当前为增加状态，则还要检查班级号是否在班级表中已经存在了，如果重号，则给出提示，如果不重号，则将数据添加到班级表中，调用 initTable()方法重新初始化表格。如果当前为修改状态，则直接更新班级表，再调用 initTable()方法重新初始化表格。

最后将 state 设置为 0，将文本框清空，并将文本框设置为不可编辑状态。

4.7 学生管理

学生管理功能与班级管理功能类似,包括学生的增、删、改三项功能,界面如图 4.3 所示。与班级管理界面不同的是,学生管理界面的上方有一个班级选择组合框,用于选择学生所在的班级。

```java
public class StudentManage extends JDialog implements ActionListener {
    private JFrame parent;
    private JComboBox cbClass;
    private JLabel lbClassName;
    private JLabel lbStudentNo;
    private JTextField tfStudentNo;
    private JLabel lbStudentName;
    private JTextField tfStudentName;
    private JLabel lbGender;
    private JComboBox cbGender;
    private JLabel lbTel;
    private JTextField tfTel;
    private JLabel lbBirthday;
    private JTextField tfBirthday;
    private JButton btOK;
    private JTable tbStudent;
    private MyTableModel model;
    private JButton btAppend;
    private JButton btEdit;
    private JButton btDelete;
    private JButton btCancel;
    private int state;     // 0. 初始状态,  1. 添加状态,  2. 修改状态
    public StudentManage(JFrame parent){
        super(parent,"学生管理",true);
        this.parent = parent;
        state = 0;
        createGUI();
    }
    private void createGUI(){
        Container c = this.getContentPane();
        c.setLayout(new BorderLayout(10,10));
        JPanel northPanel = createNorthPanel();
        c.add(northPanel,BorderLayout.NORTH);
        JPanel southPanel = createSouthPanel();
        c.add(southPanel,BorderLayout.CENTER);
        this.pack();
        this.setLocationRelativeTo(parent);
        this.setVisible(true);
    }
```

```java
private JPanel createNorthPanel(){
    JPanel classNamePanel = new JPanel(new BorderLayout());
    lbClassName = new JLabel("班级");
    ClassDao cd = new ClassDao();
    List<ClassEntity> list = cd.selectAll();
    String[] classNames = new String[list.size()];
    for(int i=0; i<list.size(); i++){
        ClassEntity ce = list.get(i);
        classNames[i] = ce.getId() + "-" + ce.getName();
    }
    cbClass = new JComboBox(classNames);
    classNamePanel.add(lbClassName,BorderLayout.WEST);
    classNamePanel.add(cbClass,BorderLayout.EAST);
    JPanel studentNoPanel = new JPanel(new BorderLayout());
    lbStudentNo = new JLabel("学号");
    tfStudentNo = new JTextField(10);
    studentNoPanel.add(lbStudentNo,BorderLayout.WEST);
    studentNoPanel.add(tfStudentNo,BorderLayout.EAST);
    JPanel studentNamePanel = new JPanel(new BorderLayout());
    lbStudentName = new JLabel("姓名");
    tfStudentName = new JTextField(10);
    studentNamePanel.add(lbStudentName,BorderLayout.WEST);
    studentNamePanel.add(tfStudentName,BorderLayout.EAST);
    JPanel genderPanel = new JPanel(new BorderLayout());
    lbGender = new JLabel("性别");
    String[] genders = {"男", "女"};
    cbGender = new JComboBox(genders);
    genderPanel.add(lbGender,BorderLayout.WEST);
    genderPanel.add(cbGender,BorderLayout.EAST);
    JPanel telPanel = new JPanel(new BorderLayout());
    lbTel = new JLabel("电话");
    tfTel = new JTextField(10);
    telPanel.add(lbTel,BorderLayout.WEST);
    telPanel.add(tfTel,BorderLayout.EAST);
    JPanel birthdayPanel = new JPanel(new BorderLayout());
    lbBirthday = new JLabel("出生日期");
    tfBirthday = new JTextField(10);
    birthdayPanel.add(lbBirthday,BorderLayout.WEST);
    birthdayPanel.add(tfBirthday,BorderLayout.EAST);
    JPanel centerPanel = new JPanel(new GridLayout(2,3,10,2));
    centerPanel.add(classNamePanel);
    centerPanel.add(studentNoPanel);
    centerPanel.add(studentNamePanel);
    centerPanel.add(genderPanel);
    centerPanel.add(telPanel);
    centerPanel.add(birthdayPanel);
```

```java
            JPanel   panel = new JPanel(new BorderLayout());
            panel.add(new JPanel(),BorderLayout.WEST);
            panel.add(new JPanel(),BorderLayout.NORTH);
            panel.add(centerPanel,BorderLayout.CENTER);
            cbClass.addActionListener(new classNameMoniter());
            cbClass.setSelectedIndex(-1);
            return panel;
    }
    private JPanel createSouthPanel(){
            tbStudent = new JTable();
            initTable();
            JScrollPane listPanel = new JScrollPane(tbStudent);
            JPanel buttonPanel = new JPanel(new GridLayout(5,1));
            btOK = new JButton("保存");
            btAppend = new JButton("增加");
            btEdit = new JButton("修改");
            btDelete = new JButton("删除");
            btCancel = new JButton("返回");
            buttonPanel.add(btOK);
            buttonPanel.add(btAppend);
            buttonPanel.add(btEdit);
            buttonPanel.add(btDelete);
            buttonPanel.add(btCancel);
            cbClass.setEditable(false);
            tfStudentNo.setEditable(false);
            tfStudentName.setEditable(false);
            cbGender.setEditable(false);
            tfTel.setEditable(false);
            tfBirthday.setEditable(false);
            btOK.addActionListener(this);
            btAppend.addActionListener(this);
            btEdit.addActionListener(this);
            btDelete.addActionListener(this);
            btCancel.addActionListener(this);
            JPanel eastPanel = new JPanel(new BorderLayout());
            eastPanel.add(buttonPanel,BorderLayout.NORTH);
            JPanel panel = new JPanel(new BorderLayout(2,0));
            panel.add(new JPanel(),BorderLayout.WEST);
            panel.add(listPanel,BorderLayout.CENTER);
            panel.add(eastPanel,BorderLayout.EAST);
            panel.add(new JPanel(),BorderLayout.SOUTH);
            return panel;
    }
    private void initTable(){
            StudentDao dao = new StudentDao();
            String selectedItem = (String) cbClass.getSelectedItem();
```

```java
            String classNo=null;
            if(selectedItem!=null){
                  String words[]= selectedItem.split("-");
                  classNo = words[0];
            }
            List<Student> list = dao.selectByClassId(classNo);
            model = createTableModel(list); //建立模板列表模型
            tbStudent.setModel(model);
            tbStudent.getColumn("班级号").setPreferredWidth(60);
            tbStudent.getColumn("学号").setPreferredWidth(60);
            tbStudent.getColumn("姓名").setPreferredWidth(60);
            tbStudent.getColumn("性别").setPreferredWidth(40);
            tbStudent.getColumn("电话").setPreferredWidth(120);
            tbStudent.getColumn("出生日期").setPreferredWidth(120);
            tbStudent.setSelectionMode(ListSelectionModel.SINGLE_SELECTION);
      }
      public MyTableModel createTableModel(List<Student> list){
            String[] strColumnName = new String[6];
            strColumnName[0] = "班级号";
            strColumnName[1] = "学号";
            strColumnName[2] = "姓名";
            strColumnName[3] = "性别";
            strColumnName[4] = "电话";
            strColumnName[5] = "出生日期";
            Object[][] tableData = new Object[list.size()][strColumnName.length];
            for(int i =0; i<tableData.length; i++){
                  Student stu = list.get(i);
                  tableData[i][0] = stu.getClassNo();
                  tableData[i][1] = stu.getId();
                  tableData[i][2] = stu.getName();
                  tableData[i][3] = stu.getGender();
                  tableData[i][4] = stu.getTel();
                  tableData[i][5] = stu.getBirthday();
            }
            return new MyTableModel(tableData, strColumnName);
      }
      class classNameMoniter implements ActionListener{
            public void actionPerformed(ActionEvent arg0) {
                  if(tbStudent!=null){
                        initTable();
                  }
            }
      }
      public void actionPerformed(ActionEvent e){
            JButton buttone = (JButton) e.getSource();
            if(buttone==btAppend){
```

```java
            if(state==1) //如果已经是添加状态，则不处理
                return;
            else if(state == 2){ //如果是编辑状态，已经编辑的内容是否作废
                int select = JOptionPane.showConfirmDialog(this, "放弃修改？");
                if(select!=JOptionPane.YES_OPTION){
                    return;
                }
            }
            tfStudentNo.setEditable(true);
            tfStudentName.setEditable(true);
            tfTel.setEditable(true);
            tfBirthday.setEditable(true);
            cbClass.setEnabled(true);
            tfStudentNo.setText("");
            tfStudentName.setText("");
            tfTel.setText("");
            tfStudentName.setText("");
            tfBirthday.setText("");
            tfStudentNo.requestFocus();
            state = 1;
        }
        else if(buttone==btEdit){
            if(state==1){
                if(JOptionPane.showConfirmDialog(this, "放弃当前数据的添加吗？")
                        !=JOptionPane.YES_OPTION){
                    return;
                }
            }
            else if(state==2){
                if(JOptionPane.showConfirmDialog(this, "放弃当前数据的修改吗？")
                        !=JOptionPane.YES_OPTION){
                    return;
                }
            }
            int row =tbStudent.getSelectedRow();
            if(row==-1){
                JOptionPane.showMessageDialog(this, "请选中一个学生！");
                return;
            }
            String classNo = (String) tbStudent.getValueAt(row,0);
            String studentNo = (String) tbStudent.getValueAt(row,1);
            String studentNname = (String) tbStudent.getValueAt(row,2);
            String studentGender = (String) tbStudent.getValueAt(row,3);
            String studentTel = (String) tbStudent.getValueAt(row,4);
            Date studentBirthday = (Date) tbStudent.getValueAt(row,5);
            tfStudentNo.setEditable(false);
```

```java
            tfStudentName.setEditable(true);
            tfTel.setEditable(true);
            tfBirthday.setEditable(true);
            cbClass.setEnabled(false);
            tfStudentNo.setText(studentNo);
            tfStudentName.setText(studentNname);
            tfTel.setText(studentTel);
            tfBirthday.setText(studentBirthday.toString());
            tfStudentName.requestFocus();
            state = 2;
    }
    else if(buttone==btDelete){
            int row =tbStudent.getSelectedRow();
            if(row==-1){ //选中一个学生才能删除
                JOptionPane.showMessageDialog(this, "请选中一个学生！");
                return;
            }
            String studentNo = (String) tbStudent.getValueAt(row,1);
            ScoreDao sd = new ScoreDao();
            List<Score> list = sd.selectByStudentId(studentNo);
            if(list.size()!=0){ //有成绩的学生不能删除
                JOptionPane.showMessageDialog(this, "该生有成绩，不能删除！");
                return;
            }
            if(JOptionPane.showConfirmDialog(this, "确认？")!=JOptionPane.YES_OPTION){
                return;
            }
            if(studentNo.equals(tfStudentNo.getText().trim())){
                if(state == 2){   //删除的恰好是正在修改的学生，清空编辑框
                    tfStudentNo.setText("");
                    tfStudentName.setText("");
                    tfTel.setText("");
                    tfBirthday.setText("");
                    tfStudentNo.requestFocus();
                    state=0;
                }
            }
            StudentDao dao = new StudentDao();
            dao.delete(studentNo);
            initTable();
    }
    else if(buttone==btOK){
            StudentDao dao = new StudentDao();
            String selectedItem = (String) cbClass.getSelectedItem();
            String classNo=null;
            if(selectedItem!=null){
```

```java
            String words[]= selectedItem.split("-");
            classNo = words[0];
        }
        else{
            JOptionPane.showMessageDialog(this, "请选择班级！");
            return;
        }
        String studentNo = tfStudentNo.getText().trim();
        String studentName = tfStudentName.getText().trim();
        String gender = (String) cbGender.getSelectedItem();
        String tel = (String) tfTel.getText().trim();
        String birthday =   tfBirthday.getText().trim();
        SimpleDateFormat sdf = new SimpleDateFormat("yyyy-MM-dd");
        Date date=null;
        try {
            date = sdf.parse(birthday);
        } catch (ParseException e1) {
            JOptionPane.showMessageDialog(this, "日期格式有误，请重输！");
            return;
        }
        if(studentNo.isEmpty() || studentName.isEmpty() || gender.isEmpty()
                            || tel.isEmpty() || birthday.isEmpty()){
            JOptionPane.showMessageDialog(this, "数据不能为空，请重输！");
            return;
        }
        if(state==1){
            Student stu = dao.selectById(tfStudentNo.getText().trim());
            if(stu!=null){
                JOptionPane.showMessageDialog(this, "学号重复，请重输！");
                return;
            }
            else{
                stu = new Student(studentNo,studentName,gender,tel,date,classNo);
                dao.insert(stu);
                initTable();
            }
        }
        else if(state==2){
            Student stu = new Student(studentNo,studentName,gender,tel,date,classNo);
            dao.update(stu);
            initTable();
        }
        state = 0;
        tfStudentNo.setText("");
        tfStudentName.setText("");
        tfTel.setText("");
```

```
                    tfBirthday.setText("");
                    tfStudentNo.requestFocus();
                    tfStudentNo.setEditable(false);
                    tfStudentName.setEditable(false);
                    tfTel.setEditable(false);
                    tfBirthday.setEditable(false);
                    tfStudentNo.requestFocus();
                }
                else if(buttone==btCancel){
                    this.dispose();
                }
            }
        }
```

StudentManage 类继承于 JDialog 类，并实现了 ActionListener 接口，方法 createGUI()、createNorthPanel()、createSouthPanel()用于创建对话框中的组件。

方法 initTable()初始化表格，使用 StudentDao 对象读取学生表，将指定班级的学生存放在 list 中，然后调用 createTableModel()方法创建表格模型，通过 JTable 的 setModel()方法将创建好的表格模型设置给 JTable 对象 tbStudent，最后设置表格每一列的宽度，并设置表格只能选中一行。

方法 createTableModel()的参数是指定班级的学生链表，在方法中首先设计好表格每列的标题 strColumnName，然后用班级链表给表格数据 tableData 赋值，最后创建 MyTableModel 对象，并将该对象作为返回值。

学生管理也是在 actionPerformed()方法中完成各个按钮的处理，各功能的实现与班级管理类似，不再重复叙述。

与班级管理不同的是，当一个学生已经有成绩记录时，该学生不可以删除。

4.8 成绩管理

4.8.1 准备工作

成绩管理包括成绩录入、成绩修改和成绩查询。由于成绩管理界面中成绩表格的特殊性，以及 JTable 类的数据编辑不是很方便，我们设计 ScoreTable 类、ScoreTableModel 类和 CreateTableModel 类。

创建包 dialog.scoreTable，在该包中创建这三个类。

1. ScoreTable 类

ScoreTable 类继承于 JTable 类，代码如下：

```
public class ScoreTable extends JTable {
    public ScoreTable(ScoreTableModel model){
        super(model);
        //最后编辑的单元格失去焦点后立即结束编辑，保证最后输入的数据不丢失
        putClientProperty("terminateEditOnFocusLost", Boolean.TRUE);
```

```java
            //只能选中一行
            setSelectionMode(ListSelectionModel.SINGLE_SELECTION);
        }
        //单击单元格后立即进入编辑状态（否则要单击两次）
        public void changeSelection(int rowIndex, int columnIndex,
                boolean toggle, boolean extend){
            super.changeSelection(rowIndex, columnIndex, toggle, extend);
            super.editCellAt(rowIndex, columnIndex, null);
        }
    }
```

ScoreTable 类主要解决两个问题：一是单元格失去焦点后结束编辑状态，保证最后输入的数据能够被保存；二是单击单元格后立即进入编辑状态。

2. ScoreTableModel 类

在成绩录入和修改界面中，成绩列是可以编辑的，而成绩查询界面的成绩列是不可以编辑的，设计 ScoreTableModel 类来实现这一功能，代码如下：

```java
public class ScoreTableModel extends DefaultTableModel {
    int type;   //1:录入修改    2:查询
    public ScoreTableModel(Object[][] objDatat, String[] columnNames, int type)
    {
        super(objDatat , columnNames);
        this.type = type;
    }
    public boolean isCellEditable(int row,int column){
        if( (type==1) && (column==2))
            return true;
        else
            return false;
    }
    public Class getColumnClass(int columnIndex){
        if(columnIndex==2){
            return Integer.class;
        }
        else{
            return String.class;
        }
    }
}
```

类中属性 type 的值为 1 表示是录入和修改，为 2 表示是查询。如果 isCellEditable()方法返回 true，表示该单元格可以编辑；返回 false，表示不可以编辑。getColumnClass()方法设置某一列的数据类型，成绩列是整型，其他列是字符串。

3. CreateTableModel 类

为了方便在不同的类中创建 ScoreTableModel 对象，专门设计一个 CreateTableModel 类，用于创建 ScoreTableModel 对象，代码如下：

```java
public class CreateTableModel {
    //type1   对话框类型1，1：录入、修改        2：查询
    //type2   对话框类型2，2：按学号查询成绩      1：其他
    public static ScoreTableModel createModel(List<Score> list, int type1,int type2){
        String[] strColumnName = new String[3];
        if(type2==1){
            strColumnName[0] = "学号";
            strColumnName[1] = "姓名";
        }
        else{
            strColumnName[0] = "课程号";
            strColumnName[1] = "课程名";
        }
        strColumnName[2] = "成绩";
        Object[][]    tableData = new Object[list.size()][3];
        if(type2==1){
            for(int i =0; i<list.size(); i++){
                Score    score = list.get(i);
                String stuNo = score.getStudentId();
                StudentDao dao = new StudentDao();
                Student stu = dao.selectById(stuNo);
                String stuName = stu.getName();
                tableData[i][0] = stuNo;
                tableData[i][1] = stuName;
                tableData[i][2] = score.getScore();
            }
        }
        else{
            for(int i =0; i<list.size(); i++){
                Score    score = list.get(i);
                String courseNo = score.getCourseId();
                CourseDao dao = new CourseDao();
                Course course = dao.selectById(courseNo);
                String courseName = course.getName();
                tableData[i][0] = courseNo;
                tableData[i][1] = courseName;
                tableData[i][2] = score.getScore();
            }
        }
        Sreturn new ScoreTableModel(tableData, strColumnName, type1);
    }
}
```

类中只有一个静态方法 createModel()，返回 ScoreTableModel 对象，其参数分别是表格中的数据 list、对话框类型 type1（录入和修改对话框值为 1，查询对话框值为 2）和对话框类型 type2（按课程查询值为 2，按其他查询值为 1）。type1 控制成绩列是否可以编辑，type2 控制前两列的内容。

4.8.2 成绩录入与修改

成绩录入与成绩修改的界面相同，与图 4.4 类似，因此我们设计一个对话框类 ScoreDialog 来实现这两个功能的界面，然后设计两个监听器类 ScoreInputMonitor 和 ScoreEditMonitor 实现成绩录入与成绩修改的具体功能。

1. ScoreDialog 类

ScoreDialog 类主要组织界面中的各个组件，代码如下：

```java
public class ScoreDialog extends JDialog{
    JFrame parent;
    JLabel lbClass;
    JLabel lbCourse;
    JComboBox cbClass;
    JComboBox cbCourse;
    JTable tbScore;
    ScoreTableModel model;
    JButton btOK;
    JButton btCancel;
    int type;   // 1.录入  2. 修改
    public ScoreDialog(JFrame parent, String title, boolean modal, int type){
        super(parent,title, modal);
        this.parent = parent;
        this.type = type;
        createGUI();
        ActionListener listener = null;
        if(type==1){
            listener =  new ScoreInputMonitor(this);
        }
        else if(type==2){
            listener =  new ScoreEditMonitor(this);
        }
        cbClass.addActionListener(listener);
        cbCourse.addActionListener(listener);
        btOK.addActionListener(listener);
        btCancel.addActionListener(listener);
        this.setVisible(true);
    }
    private void createGUI(){
        //创建对话框上半部分
```

```java
JPanel classPanel = new JPanel(new FlowLayout(FlowLayout.LEFT));
lbClass = new JLabel("班级");
ClassDao cd = new ClassDao();
List<ClassEntity> listClass = cd.selectAll();
String[] classNames = new String[listClass.size()];
for(int i=0; i<listClass.size(); i++){
    ClassEntity ce = listClass.get(i);
    classNames[i] = ce.getId() + "-" + ce.getName();
}
cbClass = new JComboBox(classNames);
cbClass.setSelectedIndex(-1);
classPanel.add(lbClass);
classPanel.add(cbClass);
JPanel coursePanel = new JPanel(new FlowLayout(FlowLayout.LEFT));
lbCourse = new JLabel("课程");
CourseDao cdao = new CourseDao();
List<Course> listCourse = cdao.selectAll();
String[] courseNames = new String[listCourse.size()];
for(int i=0; i<listCourse.size(); i++){
    Course course = listCourse.get(i);
    courseNames[i] = course.getId() + "-" + course.getName();
}
cbCourse = new JComboBox(courseNames);
cbCourse.setSelectedIndex(-1);
coursePanel.add(lbCourse);
coursePanel.add(cbCourse);
JPanel northPanel = new JPanel(new GridLayout(1,2));
northPanel.add(classPanel);
northPanel.add(coursePanel);
Container c = this.getContentPane();
c.setLayout(new BorderLayout(10,10));
c.add(northPanel, BorderLayout.NORTH);
//创建对话框下半部分
StudentDao dao = new StudentDao();
List<Score> list = new ArrayList();
ScoreTableModel model= CreateTableModel.createModel(list,1,1);
tbScore = new ScoreTable(model);
tbScore.setModel(model);
JScrollPane listPanel = new JScrollPane(tbScore);
c.add(listPanel, BorderLayout.CENTER);
btOK = new JButton("确定");
btCancel = new JButton("返回");
JPanel buttonPanel = new JPanel(new FlowLayout(FlowLayout.CENTER));
```

```java
                buttonPanel.add(btOK);
                buttonPanel.add(btCancel);
                c.add(buttonPanel,BorderLayout.SOUTH);
                this.pack();
                this.setLocationRelativeTo(parent);
            }
            String getClassNo(){
                String selectedItem = (String) cbClass.getSelectedItem();
                if(selectedItem!=null){
                    String words[] = selectedItem.split("-");
                    return words[0];
                }
                return null;
            }
            String getCourseNo(){
                String selectedItem = (String) cbCourse.getSelectedItem();
                if(selectedItem!=null){
                    String words[] = selectedItem.split("-");
                    return words[0];
                }
                return null;
            }
        }
```

方法 createGUI()创建对话框中的各个组件，并将它们组织在一起。在构造方法中调用 createGUI()方法，并注册监听器（监听器部分在后面给出，这里可以将监听器有关的代码暂时注释掉）。方法 getClassNo()和 getCourseNo()分别返回班级组合框中所选择的班级号和课程组合框中所选择的课程号。

2. ScoreInputMonitor 类

ScoreInputMonitor 类是成绩录入的监听器类，代码如下：

```java
        public class ScoreInputMonitor implements ActionListener{
            ScoreDialog sd;
            public ScoreInputMonitor(ScoreDialog sd){
                this.sd = sd;
            }
            public void actionPerformed(ActionEvent e) {
                Object obj = e.getSource();
                String classNo = sd.getClassNo();
                String courseNo = sd.getCourseNo();
                if(obj instanceof JComboBox){
                    if( (classNo!=null)&&(courseNo!=null)){
                        ScoreDao scoreDao = new ScoreDao();
                        List< Score> scoreList = scoreDao.selectByClassIdCourseId(classNo, courseNo);
```

```java
            if( (scoreList!=null) && (scoreList.size()>0) ){
                JOptionPane.showMessageDialog(sd, "成绩已录入！");
                return;
            }
            else if(sd.tbScore!=null){
                StudentDao stuDao = new StudentDao();
                List<Student> listStu = stuDao.selectByClassId(classNo);
                System.out.println(listStu.size());
                List<Score> listScore = new ArrayList<Score>();
                for(int i=0; i<listStu.size(); i++){
                    Student stu = listStu.get(i);
                    String stuNo = stu.getId();
                    String stuName = stu.getName();
                    Score score = new Score(classNo,stuNo,courseNo,0);
                    listScore.add(score);
                }
                ScoreTableModel model= CreateTableModel.createModel(listScore,1,1);
                sd.tbScore.setModel(model);
            }
        }
    }
    else if(obj instanceof JButton){
        JButton buttone = (JButton) obj;
        if(buttone==sd.btOK){
            ScoreDao dao = new ScoreDao();
            if(courseNo.isEmpty() || classNo.isEmpty()){
                JOptionPane.showMessageDialog(sd, "课号和班号不能空！");
                return;
            }
            for(int i=0; i<sd.tbScore.getRowCount(); i++){
                int s = (Integer) sd.tbScore.getValueAt(i, 2);
                String stuNo = (String) sd.tbScore.getValueAt(i, 0);
                Score score = new Score(classNo, stuNo,courseNo,s);
                dao.insert(score);
            }
            sd.dispose();
        }
        else if(buttone==sd.btCancel){
            sd.dispose();
        }
    }
}
}
```

如果监听到组合框消息，则检查所选班级和课程的成绩是否已经录入，如果已经录入，则不能重复录入，给出提示信息；否则将该班级的名单从学生表中读出，根据学生表的数据生成成绩表，为对话框中的表格提供数据。

如果监听到"确定"按钮的消息，则将界面表格中的数据添加到数据库的成绩表中。

3. ScoreEditMonitor 类

ScoreEditMonitor 类是成绩修改的监听器类，代码如下：

```java
public class ScoreEditMonitor implements ActionListener {
    ScoreDialog sd;
    public ScoreEditMonitor(ScoreDialog sd){
        this.sd = sd;
    }
    public void actionPerformed(ActionEvent e){
        Object obj = e.getSource();
        String classNo = sd.getClassNo();
        String courseNo = sd.getCourseNo();
        if(obj instanceof JComboBox){
            if( (classNo!=null)&&(courseNo!=null) ){
                ScoreDao scoreDao = new ScoreDao();
                List< Score> scoreList = scoreDao.selectByClassIdCourseId(classNo, courseNo);
                if( (scoreList==null) || (scoreList.size()==0) ){
                    JOptionPane.showMessageDialog(sd, "成绩未录入！");
                    return;
                }
            }
            if(sd.tbScore!=null){
                ScoreDao scoreDao = new ScoreDao();
                List<Score> list = scoreDao.selectByClassIdCourseId(classNo,courseNo);
                sd.model = CreateTableModel.createModel(list,1,1);
                sd.tbScore.setModel(sd.model);
            }
        }
        else if(obj instanceof JButton){
            JButton buttone = (JButton) obj;
            if(buttone==sd.btOK){
                ScoreDao dao = new ScoreDao();
                if(courseNo.isEmpty() || classNo.isEmpty()){
                    JOptionPane.showMessageDialog(sd, "课号和班号不能空！");
                    return;
                }
                for(int i=0; i<sd.tbScore.getRowCount(); i++){
                    int s = (Integer) sd.tbScore.getValueAt(i, 2);;
                    String stuNo = (String) sd.tbScore.getValueAt(i, 0);
```

```
                    Score score = new Score(classNo, stuNo,courseNo,s);
                    dao.update(score);
                }
                sd.dispose();
            }
            else if(buttone==sd.btCancel){
                sd.dispose();
            }
        }
    }
}
```

如果监听到组合框消息，则检查所选班级和课程的成绩是否已经录入，如果还没录入，则不能修改，给出提示信息；否则从成绩表中读出对应的成绩，为对话框中的表格提供数据。

如果监听到"确定"按钮的消息，则用界面表格中的成绩更新数据库成绩表中的成绩。

4.8.3 成绩查询

成绩查询有两种方式：一种是查询某门课程的所有学生成绩；另一种是查询某个学生的所有课程成绩。

1. 查询课程成绩

查询课程成绩的界面如图 4.5 所示，通过对话框上方的组合框选择课程，下方的表格显示这门课程所有学生的成绩。查询课程成绩的类是 ScoreQueryByCourse，代码如下：

```
public class ScoreQueryByCourse extends JDialog implements ActionListener {
    private JFrame parent;
    private JLabel lbCourse;
    private JComboBox cbCourse;
    private JTable tbScore;
    private ScoreTableModel model;
    private JButton btCancel;
    public ScoreQueryByCourse(JFrame parent){
        super(parent, "按课程查询成绩", true);
        this.parent = parent;
        createGUI();
    }
    private void createGUI(){
        //创建对话框上半部分
        JPanel coursePanel = new JPanel(new FlowLayout(FlowLayout.LEFT));
        lbCourse = new JLabel("课程");
        CourseDao cdao = new CourseDao();
        List<Course> listCourse = cdao.selectAll();
        String[] courseNames = new String[listCourse.size()];
        for(int i=0; i<listCourse.size(); i++){
            Course course = listCourse.get(i);
```

```java
            courseNames[i] = course.getId() + "-" + course.getName();
        }
        cbCourse = new JComboBox(courseNames);
        cbCourse.setSelectedIndex(-1);
        cbCourse.addActionListener(this);
        coursePanel.add(lbCourse);
        coursePanel.add(cbCourse);
        Container c = this.getContentPane();
        c.setLayout(new BorderLayout(10,10));
        c.add(coursePanel, BorderLayout.NORTH);
        //创建对话框下半部分
        ScoreDao dao = new ScoreDao();
        List<Score> list = new ArrayList<Score>();
        ScoreTableModel model= CreateTableModel.createModel(list,2, 1);
        tbScore = new ScoreTable(model);
        tbScore.setModel(model);
        JScrollPane listPanel = new JScrollPane(tbScore);
        c.add(listPanel, BorderLayout.CENTER);
        btCancel = new JButton("返回");
        JPanel buttonPanel = new JPanel(new FlowLayout(FlowLayout.CENTER));
        buttonPanel.add(btCancel);
        c.add(buttonPanel,BorderLayout.SOUTH);
        btCancel.addActionListener(this);
        this.pack();
        this.setLocationRelativeTo(parent);
        this.setVisible(true);
    }
    public void actionPerformed(ActionEvent e){
        Object obj = e.getSource();
        if(obj instanceof JComboBox){
            String slectedItem = (String) cbCourse.getSelectedItem();
            String courseNo = null;
            if(slectedItem!=null){
                String[] words = slectedItem.split("-");
                courseNo = words[0];
            }
            if( (tbScore!=null) && (courseNo!=null)&& (!courseNo.isEmpty()) ){
                ScoreDao scoreDao = new ScoreDao();
                List< Score> scoreList = scoreDao.selectByCourseId( courseNo);
                model =   CreateTableModel.createModel(scoreList,2,1);
                tbScore.setModel(model);
            }
        }
```

```java
        else if(obj instanceof JButton){
            this.dispose();
        }
    }
}
```

createGUI()方法创建对话框中的组件,并完成注册监听器。actionPerformed()方法实现对不同事件的处理,如果是组合框事件,则根据所选课程的课程号,在成绩表中查询该门课程的所有成绩,为对话框下方的表格提供数据。如果是按钮事件,则直接退出对话框。

2. 查询学生成绩

查询学生成绩的界面如图4.6所示,通过对话框上方的组合框选择学生,下方的表格显示该学生的所有成绩。查询学生成绩的类是 ScoreQueryByStudent,代码如下:

```java
public class ScoreQueryByStudent extends JDialog implements ActionListener {
    private JFrame parent;
    private JLabel lbStudent;
    private JComboBox cbStudent;
    private JTable tbScore;
    private ScoreTableModel model;
    private JButton btCancel;
    public ScoreQueryByStudent(JFrame parent){
        super(parent, "按学生查询成绩", true);
        this.parent = parent;
        createGUI();
    }
    private void createGUI(){
        //创建对话框上半部分
        JPanel studentPanel = new JPanel(new FlowLayout(FlowLayout.LEFT));
        lbStudent = new JLabel("姓名");
        StudentDao stuDao = new StudentDao();
        List<Student> stuList = stuDao.selectAll();
        String[] stuNames = new String[stuList.size()];
        for(int i=0; i<stuList.size(); i++){
            Student stu = stuList.get(i);
            stuNames[i] = stu.getId() +"-" + stu.getName();
        }
        cbStudent = new JComboBox(stuNames);
        cbStudent.setSelectedIndex(-1);
        cbStudent.addActionListener(this);
        studentPanel.add(lbStudent);
        studentPanel.add(cbStudent);
        Container c = this.getContentPane();
        c.setLayout(new BorderLayout(10,10));
        c.add(studentPanel, BorderLayout.NORTH);
```

```java
        //创建对话框下半部分
        ScoreDao dao = new ScoreDao();
        List<Score>    list = new ArrayList<Score>();
        ScoreTableModel model= CreateTableModel.createModel(list,2, 2);
        tbScore = new ScoreTable(model);
        tbScore.setModel(model);
        JScrollPane listPanel = new JScrollPane(tbScore);
        c.add(listPanel, BorderLayout.CENTER);
        btCancel = new JButton("返回");
        JPanel buttonPanel = new JPanel(new FlowLayout(FlowLayout.CENTER));
        buttonPanel.add(btCancel);
        c.add(buttonPanel,BorderLayout.SOUTH);
        btCancel.addActionListener(this);
        this.pack();
        this.setLocationRelativeTo(parent);
        this.setVisible(true);
    }
    public void actionPerformed(ActionEvent e){
        Object obj = e.getSource();
        if(obj instanceof JComboBox){
            String selectedItem = (String) cbStudent.getSelectedItem();
            String stuNo=null;
            if(selectedItem!=null){
                String[] words = selectedItem.split("-");
                stuNo = words[0];
            }
            if( (tbScore!=null) && (stuNo!=null)&& (!stuNo.isEmpty()) ){
                ScoreDao scoreDao = new ScoreDao();
                List< Score> scoreList = scoreDao.selectByStudentId( stuNo);
                model =  CreateTableModel.createModel(scoreList,2,2);
                tbScore.setModel(model);
            }
        }
        else if(obj instanceof JButton){
            this.dispose();
        }
    }
}
```

查询学生成绩的界面与查询课程成绩的界面类似，只是将课程选择组合框换成了学生选择组合框。功能实现也类似。在监听组合框事件的代码中，根据选择学生的学号，在成绩表中查询成绩，并显示在对话框下方的表格中。

4.9 作业

1. 根据班级管理模块的设计完成课程管理模块。
2. 在查询学生成绩时，如果学生较多，直接从组合框中选择学生不是很方便，可以在查询界面中再添加一个班级选择组合框，先选择班级，然后在学生组合框中只显示该班级的学生，请编写相关代码实现这一功能。
3. 有时需要查询一个学生某门课的成绩，请设计实现这一功能。

参考文献

[1] 李明，吴琼，韩旭明，等．Java程序设计案例教程[M]．北京：清华大学出版社，2013．
[2] 朱福喜，黄昊．Java项目开发与毕业设计指导[M]．北京：清华大学出版社，2008．
[3] 黄晓东．Java课程设计案例精编[M]．2版．北京：中国水利水电出版社，2008．
[4] 裴博文．五子棋人工智能权重估值算法[J]．电脑编程技巧与维护，2008（6）：69-75．
[5] 邢森．五子棋智能博弈的研究与设计[J]．电脑知识与技术，2010，06（5）：3497-3498．